Edward Ernest Green

The Coccidae of Ceylon

Part I

Edward Ernest Green

The Coccidae of Ceylon
Part I

ISBN/EAN: 9783337237325

Printed in Europe, USA, Canada, Australia, Japan

Cover: Foto ©berggeist007 / pixelio.de

More available books at **www.hansebooks.com**

CONTENTS OF PART I.

	PAGE
PREFACE	iii
GLOSSARY OF TERMS	vii
CHAPTER I. INTRODUCTORY	1
" II. COLLECTION AND PREPARATION	8
" III. CHARACTERS AND CLASSIFICATION	14
Synopsis of Sub-families	16
" IV. CONCHASPINÆ	19
Conchaspis socialis.	20
" V. DIASPINÆ	24
Synopsis of Genera	37

Aspidiotus	*Page* 39	Aonidia loranthi	*Page* 74
" trilobitiformis	41	" obscura	76
" ficus	43	Mytilaspis	77
" rossi	45	" citricola	78
" osbeckiæ	47	" cocculi	81
" lataniæ	49	" gloverii	83
" cyanophylli	51	" gloverii *var.*	
" excisus	53	pallida	85
" putearius	54	Diaspis	86
" occultus	56	" amygdali	87
" aurantii	58	" fagrææ	91
" camelliæ	60	Fiorinia	93
" cydoniæ	62	" fioriniæ	94
" secretus	64	" saprosmæ	96
" inusitatus	66	" similis	98
Aonidia	68	" scrobicularum	100
" corniger	69	" secreta	102
" bullata	72		

THE
COCCIDÆ OF CEYLON.

BY

E. ERNEST GREEN, F.E.S.

PART I.

WITH THIRTY-THREE PLATES.

LONDON:
DULAU & CO.
1896.

PREFACE.

THE present work has been written with a twofold object. Firstly, to give a scientific and formal description of this little-studied but important group of insects; and, secondly, to enable planters and agriculturists in general to recognise these destructive pests, to understand their habits, and to learn how best to deal with them.

In the first place I have fully described all such species as have already been recorded from Ceylon, and have added descriptions of a very large number of new and hitherto unrecognised species collected in the island during the last five years. To show the proportion of these additions, I may mention that in 1891, when Mr. W. F. Kirby published his paper* on the Hemiptera of Ceylon, only seven species had been recorded. In November, 1894, at the request of several entomologists interested in the subject, I drew up a preliminary catalogue† of the species that I had then found in Ceylon. In this list, which was delayed in publication, and appeared only this year (1896), I enumerated seventy-two distinct species. This large number will be almost doubled in the present work, and, when other parts of the island have been properly explored, it is probable that considerably over two hundred species will be recognised.

To secure uniformity of terms, every species, whether new or

---

* 'Catalogue of the described Hemiptera, Heteroptera, and Homoptera of Ceylon.' W. F. Kirby, F.L.S., F.E.S. *Journal of the Linnean Society (Zoology)*, Vol. XXIV. pp. 72—176. 1891.

† 'Catalogue of Coccidæ, collected in Ceylon by Mr. E. E. Green.' *Indian Museum Notes*, Vol. IV. No. 1, 1896.

not, has been fully described according to a uniform plan. A glossary (made very full for the benefit of those unacquainted with entomological terms) is appended.

The descriptions and figures (unless otherwise stated), have been drawn up from and refer to Ceylon specimens only, and may possibly differ slightly from typical examples in such minor points as size and colour.

On each plate will be found one figure representing the insect of its natural size, in its natural position on the food plant. The other figures are not drawn to scale, but have been each enlarged to such a degree as to most conveniently show the requisite detail. The actual measurements of each species are given in the descriptive letterpress.

As the metric system is being universally adopted for scientific purposes, I have given all measurements in millimetres (mm.). For the benefit of those who are more conversant with the English standard of inches (") and lines ('''), I have shown a scale of the two systems, by which they may be compared. For a rough comparison, twenty-five millimetres might be taken as representing one inch (the actual figure would be nearer twenty-five and a half millimetres); one hundredth of an inch would thus represent one-quarter of a millimetre (0·25 mm.); or one-quarter of an inch would equal six and a quarter millimetres (6·25 mm.). But for ordinary purposes a glance at the figure on the plate representing the natural size of the insect will give the best idea of its proportions.

Signoret's classical work, the *Essai sur les Cochenilles*, and Targioni-Tozetti's Italian papers, amongst the older writers; Mr. W. M. Maskell's many important papers extending through a long series of the *Transactions of the New Zealand Institute*, and his volume on the *Scale Insects of New Zealand;* Professor Comstock's first and second Reports on Scale Insects; are all indispensable to a student of the Coccidæ. The scattered papers of Mr. J. W. Douglas and Mr. R. Newstead in the *Entomologist's Monthly*

Magazine; of the late Dr. Riley, Dr. L. O. Howard, and Professor Cockerell, in the various American journals; of Professor W. W. Froggatt in New South Wales; and Herr Karl Sulc, in Bohemia, must all be consulted, together with the writings of many other authors who have dealt with this subject.

The literature relating to the *Coccidæ* is, unfortunately, very scattered, which adds very considerably to the labour of research. I do not propose to give a complete bibliography of the subject, as a list* of the principal works has already been carefully compiled by Mr. W. M. Maskell, of Wellington, New Zealand, who has added so much to our knowledge of this group of insects by his long-continued researches on the Coccidæ of Australasia. I must, however, specially draw attention to the most admirable and exhaustive papers on Italian Coccidæ, by Professor Antonio Berlese,† whose exquisite figures and careful work will prove of great value to the morphologist and physiologist. Professor Cockerell's recently compiled 'Check List' ‡ is a very handy and useful catalogue of all the known species of Coccidæ.

My thanks are particularly due to Mr. Maskell for his unremitting kindness and assistance in determining specimens; to Professors Comstock, Howard, and Cockerell, in America; to Messrs. J. W. Douglas and Robert Newstead, in England; to Herr Sulc, in Bohemia; to Dr. W. W. Froggatt, in New South Wales; and to Mr. C. B. Lounsbury, now acting as Government Entomologist in Cape Colony; from all of whom I have received valuable papers and numerous typical specimens for comparison and study.

Dr. L. O. Howard has kindly named for me various hymenopterous parasites that prey upon Coccids in Ceylon.

---

* *Transactions of the New Zealand Institute*, 1891, p. 7.

† *Le Cocciniglie Italiane viventi sugli Agrumi*, by Dr. Antonio Berlese. Parts I. (1893), II. (1894), and III. (1896).

‡ 'Check List of the Coccidæ,' by T. D. A. Cockerell. *Bulletin of the Illinois State Laboratory of National History*, Vol. IV. pp. 318—339, 1896.

Most of the species described in the present work were collected by myself, chiefly in the neighbourhood of Punduloya at an altitude varying between three thousand and four thousand feet. Some particularly interesting species were discovered and sent to me by that most industrious entomologist, Mr. John Pole, from Hambantota, Tangalle, and Chilaw. Mr. W. D. Holland, of Balangoda, has also very kindly sent me specimens from that district.

To Professor G. B. Howes, of the Royal College of Science, South Kensington, I must tender my sincere thanks for the kindest encouragement and advice in the difficult matter of arranging for the publication of a work of this nature; and to Mr. F. Justen, of Messrs. Dulau & Co., for his very kind personal interest and care in the actual work of publication.

The lithographic plates, reproduced from my own drawings, have been most carefully printed in colours by P. W. M. Trap, of Leiden.

The chapter dealing with the economic side of the question has been compiled from various sources, and is intended to give a *résumé* of all the known methods of dealing with this group of insect pests. I have particularly made free use of the numerous valuable bulletins and reports issued by the United States Department of Agriculture, the Entomological Division of which—formerly under the guidance of the late Dr. C. V. Riley, whose early death is greatly to be deplored, and now worthily carried on by Dr. L. O. Howard—is far ahead of that of any other country in the attention paid to the cure and prevention of insect pests. This chapter will be issued as an appendix, and will appear with Part II. of the work.

E. ERNEST GREEN.

LONDON,
September 30, 1896.

PROVISIONAL GLOSSARY OF TERMS USED IN PART I.

(The following definitions relate only to the terms as applied in the description of Coccidæ. In general entomology they would have, in many cases, a much wider application.)

Abdomen.—All the hinder part of the insect posterior to the 'thorax.' The third of the three main divisions of the body (head, thorax, and abdomen).

Anal lobes; Anal plates.—A pair of small triangular hinged processes forming a valve which covers the 'anal orifice' in the *Lecaniinæ*.

Anal orifice.—The external opening of the intestine.

Anal ring.—A circumscribed chitinous ring encircling the 'anal orifice' in many Coccidæ.

Anal tubercles.—A pair of prominent rounded or conical processes situate one on each side of the 'anal orifice' in *Dactylopiinæ* and larval *Hemicoccinæ*.

Antennæ.—A pair of jointed organs or 'feelers' situated on the head. (In the *Diaspinæ* they are well developed in the male insect, but rudimentary in the female.)

Antennal.—Relating to the 'antennæ.'

Appendages.—A general term for the antennæ, mouth-parts, and limbs of an insect.

Apodema.—A conspicuous transverse band crossing the thorax in front of the 'scutellum' in male Coccidæ. (*Pl.* 11. *fig.* 2, *g.*)

Apodous.—Without legs.

Apterous.—Without wings.

Balancers.—(See 'halteres')

Biarticulate.—With two joints.

Bicuspid.—Having two points or prominences.

Carina.—A keel or ridge.

Carinated.—Keeled, ridged, or ribbed.

Castaneous.—Of the colour of a chestnut. Shining reddish brown.

Caudad.—Situated towards the tail or caudal extremity, in relation to some other part.

Caudal.—Pertaining to the tail or posterior extremity.

Cephalic.—Pertaining to the head.

Cephalad.—Towards the head—in relation to some other part.

Cephalothorax.—The anterior part of the body, comprising the head and the thorax, which in the females of *Diaspinæ* have no distinct line of separation.

Chitin.—A horny substance present in the skin and harder parts of insects.

Chitinised.—Hardened by the deposition of 'chitin.'

Chitinous.—Consisting of 'chitin.'

Circumgenital glands.—Small circular glands disposed in distinct groups round the 'genital orifice.' (Sometimes termed 'grouped glands.') (*Pl.* 1. *fig.* 14, *a, b, c.*)

Compressed.—Flattened from side to side, as opposed to 'depressed' (which see).

Contiguous.—Touching.

Explanation of Terms.

Cornea.—The horny convex covering of the eye.
Coxa.—The basal joint of the leg. (*Pl.* II. *fig.* 6, *a*.)
Cuticle.—The thin outer skin of the leaves and other parts of a plant.
Depressed.—Flattened from above downwards, as opposed to 'compressed' (which see).
Diagnosis.—A short distinctive description by which the genus, species, &c., of the insect may be recognised.
Digitules.—Appendages frequently present on the feet of Coccidæ, either broad and spatulate, or in the form of knobbed hairs. (*Pl.* II. *fig.* 6, *g*, *h*.)
Dimerous.—Composed of two pieces.
Dorsad.—Towards the 'dorsum.' (A term of comparative direction.)
Dorsal.—Relating to the back or upper parts of the body.
Dorsal scale.—The part of the covering scale (puparium) of the *Diaspinæ* that lies above the insect, as opposed to the 'ventral scale' which completes the puparium below.
Dorsum.—The back or upper parts of the body.
Eccentric.—Away from the centre. Out of centre.
Ecdysis.—The periodical 'moult' or change of skin.
Emarginate.—Having a notch as if a piece had been cut out.
Exuviæ.—The discarded skins shed at the periodical moults (ecdyses).
Femur.—The thigh or upper part of the leg, situate between the 'trochanter' and the 'tibia.' (For purposes of measurement, the 'trochanter' and 'femur,' being fused together, are considered as one piece.) (*Pl.* II. *fig.* 6, *c*.)
Filiform.—Thread-like.
Fimbriate, fimbriated.—Fringed. With finely divided margin.
Function.—The action or operation of any 'organ' (which see).
Funicle.—The long terminal joint of the antennæ of larval *Diaspinæ*.
Genæ.—The cheeks. The sides of the head behind the eyes. (*Pl.* II. *fig.* 4, *c*.)
Genital spike.—The sheath of the penis, which in the males of *Diaspinæ* takes the form of a long mucronate spike. (*Pl.* II. *fig.* 10.)
Gestation.—The period during which the gravid female is maturing the ova or embryos.
Grouped glands.—(See 'Circumgenital glands.')
Halteres.—A pair of small organs (sometimes called 'Balancers') which replace the hind wings in the males of Coccidæ and the two-winged flies (Diptera). (In the Coccidæ they take the form of a strap-shaped basal part, with one or more longish, stout-hooked bristles on the extremity.) (*Pl.* II. *fig.* 3.)
Homologous.—An organ or any part of an animal is said to be 'homologous' with another part when the two have the same origin—without of necessity having the same function. (As opposed to 'analogous,' in which the two parts have a similar function, although with a different origin.)
Honey-dew.—A sweet, viscid substance excreted by Coccidæ and some other homopterous insects.
Incised.—With marginal slits or notches.
Laterad.—Towards the side. (As indicating the position of one part in relation to another.)

Larva, larval stages.—The immature insect. The early stages of the insect previous to the pupa. (In the *Diaspinæ* the larval stages end with the second moult.)

Line (′′′).—The twelfth part of an inch.

Lobe.—Any prominent rounded process. More especially applied to the rounded, tooth-like processes on the margin of the 'pygidium' in the *Diaspinæ*. Sometimes also applied to the prominent lateral expansions of the abdominal segments.

Mentum.—The lower part of the mouth, which in the Coccidæ takes the form of a conical process channelled on its upper surface to receive the rostral setæ or sucking-tube.

Mesad.—Situated towards the middle—in relation to some other part.

Mesal.—Relating to the middle.

Mesosternum.—The ventral parts of the 'mesothorax.'

Mesothorax.—The median division of the thorax, bearing the second pair of legs and the fore wings—when present.

Metamorphosis.—A change of form. The transformations of an insect during its development.

Metasternum.—The ventral parts of the 'metathorax.'

Metathorax.—The hinder division of the thorax, bearing the third pair of legs and the hind wings—when present (or the 'halteres in the males of the Coccidæ).

Millimetre (mm).—The 1000th part of a metre. Approximately equal to the twenty-fifth part of an inch.

Monomerous.—Of a single piece or joint.

Mucronate.—Sharply pointed.

Nervures.—The so-called veins of the wing.

Ocelli.—The simple or supplementary eyes. (In the greater number of male Coccidæ the 'ocelli' are greatly enlarged, and take the place of the true eyes which, in such cases, are quite rudimentary.)

Œsophagus.—That part of the alimentary canal connecting the mouth with the stomach.

Organ.—Any part of the body concerned in some action or function.

Oviparous.—Producing eggs.

Oviposition.—The act of laying eggs.

Ovoviviparous.—Producing eggs which are hatched within the body of the parent or during the process of extrusion.

Parasitised.—Containing parasites. Affected or attacked by parasites.

Parastigmatic glands.—Small circular glands sometimes present round the openings of the spiracles. (They secrete a waxy powder similar to that produced by the 'circumgenital glands.')

Parthenogenesis.—Reproduction without the assistance of the male by a process of internal budding, by which several or many successive generations of fertile females may be produced.

Pellicles.—The 'exuviæ' or cast larval skins. More particularly applied to the hardened larval skins attached to the 'puparia' of the *Diaspinæ*.

Plate.—Any broad flattened piece. Definite horny tracts of the tegument.

Processes.—Any prominent portions of the body not otherwise definable.

Prosternum.—The ventral parts of the 'prothorax.'

Prothorax.—The anterior division of the thorax, bearing the first pair of legs.
Puparium.—Used here for the covering-scale formed by the *Diaspinæ*.
Pupa.—The 'chrysalis' or resting stage of an insect.
Pupiform.—Shaped like a 'pupa' or 'chrysalis.'
Pygidium.—The compound terminal segment of the *Diaspinæ* and *Conchaspinæ*.
Reniform.—Shaped like a kidney.
Rostral apparatus.—The mouth-parts, comprising (in the Coccidæ) the 'rostrum,' 'mentum,' and 'rostral setæ.'
Rostral setæ.—The four long hair-like processes which together form the sucking-tube.
Rostrum.—Used here for the upper parts of the mouth, from which spring the 'rostral setæ;' probably consisting of the clipeus and labrum fused together.
Rugose.—With fine wrinkled lines.
Sac.—The separate cottony envelope secreted by many Coccidæ.
Scale.—The puparium of a Diaspid. The waxy covering of a male Lecaniid. Used also as a general term or abbreviation for 'scale-bugs' or 'scale insects.'
Scale-bug, Scale insect.—Popular terms for any member of the family Coccidæ.
Scutes.—Circumscribed chitinous plates on the surface of the body, particularly on the several parts of the thorax.
Scutellum.—A conspicuous shield-shaped 'scute' on the dorsal surface of the 'metathorax.'
Secretion.—Matter produced by the various glands of the body. More particularly the waxy, fibrous, cottony, or silken substances of which the coverings of various Coccidæ are composed.
Secretionary.—Consisting of 'secretion.'
Secretionary supplement.—That part of a Diaspid scale extending beyond or around the 'pellicles.'
Secretory.—Concerned in the process of 'secretion.'
Segments, Somites.—The transverse divisions of the body.
Serrated.—With margin notched like a saw.
Serratulate.—Having very fine saw-like notches.
Seta.—A stout hair or bristle.
Setiferous.—Bearing 'setæ.'
Somites.—(See 'Segments.')
Spatulate.—Shaped like a spatula. Flattened and dilated at the tip.
Spinnerets.—Organs concerned in the emission of the silky or cottony filaments of which the 'scales' or 'sacs' of various Coccidæ are composed.
Spiracles, Stigmata.—The respiratory orifices. (*Pl.* I., *fig.* 10, *d, f.*)
Squames.—The flattened, fimbriated, or spine-like marginal processes of the 'pygidium' in the *Diaspinæ*, other than the 'lobes' and true spines. ('Plates' of Comstock; 'scaly hairs' of Maskell). (*Pl.* I. *fig.* 14, *k, l.*)
Stigmata.—(See 'Spiracles.')
Sub.—As a prefix, indicating approximateness.
Sub-circular.—Approximately circular. Not quite circular.
Suctorial.—'Suctorial insects,' are such as have mouths constructed for sucking and take their food in a liquid form; in contradistinction to 'mandibulate insects' which have jaws and bite their food.

Explanation of Terms.

Synopsis.—A tabulated arrangement showing at a glance the distinctive characters of the several families, genera, or species under consideration.

Tarsus.—The terminal joints of the leg, succeeding the 'tibia.' (*Pl.* II. *fig.* 6, *e.*)

Tarsal.—Belonging to the 'tarsus.'

Test.—The secretionary covering of various Coccidæ; especially such as are of a waxy, horny, or glassy consistence.

Thorax.—The second of the main divisions of the body; that part that bears the legs and wings (when present).

Thoracic.—Belonging to the 'thorax.'

Tibia.—The single joint of the leg immediately succeeding the 'femur' and preceding the 'tarsus.' (*Pl.* II. *fig.* 6, *d.*)

Triarticulate.—With three joints.

Tricuspid.—With three points.

Trimerous.—In three parts or joints.

Trochanter.—The small joint connecting the 'femur' with the 'coxa,' and usually firmly fused with the former. (*Pl.* II. *fig.* 6, *b.*)

Truncate.—With extremity having the appearance of being abruptly cut off.

Tumescent.—Swollen.

Ungual.—Belonging to the claw.

Ventral.—Relating to the under surface of the body.

Ventral scale.—The under part of the 'puparium' in the *Diaspinæ*, interposed between the insect and the plant.

Wax glands.—Small circular glands concerned in the secretion of waxy matter: present on the pygidium as 'circumgenital glands,' and round the 'spiracles' as 'parastigmatic glands; occurring also on other parts of the body in various families.

SCALE OF INCHES AND MILLIMETRES.

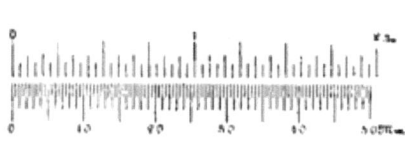

THE COCCIDÆ OF CEYLON.

CHAPTER I.

INTRODUCTORY.

THE Coccidæ, or 'Scale Insects,' subsist upon vegetable sap, pumping it up by means of a proboscis or long hair-like tube which they insert deep into the tissues of the plant. They may all be classed as either actual or potential insect pests. Many of them are at present rare and local in Ceylon; long may they remain so! Others, unfortunately, are amongst our worst enemies. I need only mention the 'Green-bug' (*Lecanium viride*), that worked such havoc in our coffee plantations some ten years ago and still remains with us. In the earlier days of coffee, the 'Black-bug' (*Lecanium nigrum*) and 'Brown-bug' (*Lecanium coffeæ*) were only a little less destructive; and the latter is still sometimes very troublesome on tea plants. At the present time a small insect, known in America as the 'Greedy Scale' (*Aspidiotus camelliæ*), has found its way into our tea estates, and is responsible for many weakly plants. There is scarcely a single cultivated plant that is not subject to the attacks of one or more species of Scale Insects; some few plants seem particularly attractive to these pests. The common guava tree, for instance, is never free from 'bug.' On one small tree of this kind I have counted as many as seven distinct species at one time, and such a tree growing in the midst of a field of tea or coffee will form a stronghold for such pests, and a source of infection, unless speedily eradicated. Ornamental plants — especially those grown in sheltered places, such as palms and ferns in pots — are particularly subject to attack. It will often be noticed that a creeper growing under the eaves of a house may be thickly covered with scale, while a similar plant grown in the open will be quite free. Similarly, a coffee or tea bush, sheltered by some overhanging rock, will usually be more subject to 'Red-spider,' 'Mealy-bug,' and 'Scale,' than are more exposed trees.

It is a curious fact, in connexion with Scale Insects, that particular species are liable—under certain circumstances, such as the accidental extermination or reduction of some natural enemy—to suddenly spring into prominence. Another great source of danger is the introduction of new species from other countries. This may easily happen with the importation of foreign fruit and growing plants. I have myself seen living specimens of the mussel-scale' (*Mytilaspis pomorum*) upon Tasmanian apples sold in Ceylon. In California and some other American states a special quarantine officer is appointed, whose duty is to examine all importations of plants and fruit, and to disinfect or destroy any infected stock. It is extremely probable that the Green-bug (*Lecanium viride*) was an introduction, though its original home has never been determined. The Orthezia (*O. insignis*) that suddenly appeared in the Botanical Gardens, Peradeniya, was certainly brought into the country with living plants.

To show the exceptional danger of introduced pests, I cannot do better than quote from the very excellent report for the year 1895 of the Government Entomologist (Mr. C. P. Lounsbury), at the Cape of Good Hope. On page 14 he writes:—

'The increased destructiveness of imported insects is well known. Few insects in this country now demand greater attention than the Phylloxera of the Vine. In the Eastern United States, the original home of this insect, it is rarely heard of as injurious, but, throughout those countries into which it has been unfortunately introduced, its terrible destructiveness, under new and more favourable conditions, has manifested itself. Another almost equally striking illustration in Cape Colony is the Australian bug (*Icerya purchasi*, Mask.). In Australia its ravages are never very extensive, but its depredations in this country will not soon be forgotten. The increased ravages of injurious insects in new countries result from the improved conditions under which they become placed. The relations which have existed, perhaps for centuries, between them and their food plants, their parasites and predacious enemies, such as other insects and birds, are all suddenly broken. In their importation their natural enemies, which had previously preserved a balance between them and the vegetable world, are all left behind, and they are free to increase and multiply without hindrance or molestation, until native parasites and predacious insects and animals acquire the habit of preying upon them. By the introduction of some of their enemies in the land from which they came, man can in exceptional cases do much towards counteracting their increase, but in the vast majority of cases generations are required before a proper balance becomes again established.'

As some little set-off against the destructiveness of many of the Scale Insects, a few species may be quoted that are of economic use. The well-known Cochineal insect (*Coccus cacti*), producing the red colouring matter known as Cochineal, is a case in point. Another species (*Tachardia lacca*), secretes a resinous substance from which is made the 'lac' or 'shellac' of commerce, while from the insect itself is prepared the fine crimson pigment known as 'lake.' *Ericerus pela*, a Chinese insect, secretes copiously a waxy matter that is used in the manufacture of candles in that country.

We have in Ceylon representatives of all these types. A Cochineal insect, identical with or closely allied to the *Coccus cacti*, breeds upon the wild cactus plant (*Opuntia Dillenii*); but I do not know if it has ever been locally utilised or cultivated. There are two species of *Tachardia* producing 'lac,' the product of which is collected by the natives and used in the manufacture of varnishes. *Ceroplastes ceriferus* is an abundant producer of insect wax, but the quality is said to be unsuitable for economic purposes.

The absence of any winter or resting stage for plant life in Ceylon allows of the continuous activity of insect pests. In countries where a dead season prevails for a part of the year the Scale Insects produce only a limited number of broods, varying in different species, before the fall of leaf and cessation of the flow of sap compels them to cease their depredations. The greater number of the insects die off, while the balance pass the winter either in a dormant state or in the egg stage. They are now subject to many dangers that tend to still further reduce their numbers. The bareness of the branches lays them open to attack from insectivorous birds that pry into every crank and cranny in search of their accustomed food. Abnormal wet or frosts will pick off the more exposed individuals, and but a comparatively small number will survive to continue the species. Hence, in England, we seldom if ever hear of any such serious plagues as occur in more southern countries. In Ceylon, however, where perpetual summer reigns, an endless succession of broods follows one upon another, only partially checked by the season of heavy rains. In the majority of cases the insects may be found in every month of the year.

It is during the rainy seasons, or rather immediately after that period, that we are able to take our Coccid enemies at a disadvantage. Their numbers have been reduced by a fungoid disease to which they are specially subject at this season, and the survivors are probably weakened from the same cause. It is now that remedial measures can be best undertaken and insecticides will have the best effect. An account of the most approved methods of treatment will be found in an appendix devoted to the subject.

Another fact that cannot fail to attract the attention of a student of the Coccidæ is the wide distribution of many of the species. This is especially the case with those affecting fruit trees. Looking through our Ceylon list, we find individual species that occur in all quarters of the globe. *Mytilaspis citricola*, for instance, will be found wherever plants of the orange and citron family are cultivated, while *Diaspis amygdali* attacks fruit trees of several kinds in such widely separate countries as America, South Africa, Ceylon, Japan, and Australia. The reason of this wide distribution is not difficult to understand when we remember the ease with which living Coccids may be transported with fruit and growing plants. Year by year the commoner species of Scale Insects are becoming more cosmopolitan.

Every tree, shrub, or plant, would soon be completely overrun with 'Scale-bugs,' and other insect pests, if it were not for the good services of numerous natural enemies, which may be primarily divided into the two classes of vegetable and animal. Under the first class will fall the mould-like fungus that attacks and destroys many Scale Insects during the wet season. In the second class are the many insects that prey upon Coccidæ and their allies. The predatory insects, again, belong to two distinct categories, external feeders and internal feeders.

The 'Lady-bird' beetles are the principal agents among the external feeders. Many species of these useful little insects live entirely upon Scale-bugs, devouring them with avidity and sometimes entirely freeing a tree of these pests. The process may often be observed where a colony of bugs has overrun some tree trunk. Many of the individuals are seen to be mere empty shells, each with a jagged hole in the back. Further on, one of the little beetles, or its elongated alligator-shaped grub, may perhaps be seen at work, greedily tearing open and devouring its defenceless prey, which is unable to escape or to make any movement. Unfortunately, our native Lady-birds have enemies of their own, and are consequently greatly handicapped in their good work. But grand results have followed, in other countries, the introduction of a foreign beetle which, having left its own enemies behind, finds itself free to increase without check so long as suitable food is forthcoming. The introduction of the 'Vedalia beetle' into California is a well-known case in question, resulting as it did in the complete clearance of the dreaded 'Fluted Scale' (*Icerya purchasi*) from the orange plantations of that country.

Another useful friend to the planter is the larva of a species of 'Lace-wing' fly. This little animal is provided with long sickle-shaped jaws, hollow and perforated at their tips, which it plunges into the body of its victim and through which it sucks up the juices of the unfortunate insect, whose empty skin is then fastened upon the back of the devourer together with those of former sufferers.

The caterpillars of butterflies and moths are usually themselves vegetable feeders, and often very injurious to plants. But we have in Ceylon at least one of each class that has developed carnivorous tastes and adopted a diet of Scale Insects. The larva of a small blackish butterfly (*Spalgis epius*) feeds entirely upon Mealy-bugs; and that of a small noctuid moth (*Eublemma cocciphaga*, Hampsn), preys upon several kinds of Scale-bugs, forming a neat covering for its body out of their empty scales.

The internal feeders are the hymenopterous parasites—that vast army of minute wasp-like insects (belonging chiefly to the *Chalcididae*), that insinuate their eggs into and pass their earlier stages within the bodies of their victims. Their small size may be realised when we find that four or five of these little wasps may be developing simultaneously within the body of a Scale Insect less than one eighth of an inch in length. The presence of one or more of these parasites does not always seem to deprive the insect of its reproductive powers; but such diseased individuals cannot be so prolific as healthier insects, and many of them are certainly killed off before attaining the adult stage. I have sometimes found a colony of Scale-bugs in which nearly every individual had been parasitised.

It was at one time supposed by planters that the ants which are so constantly in attendance upon Scale-bugs were preying upon them, and the formidable 'Red-ant' (*Œcophylla smaragdina*) was actually imported into some estates with the view of exterminating Black-bug on the coffee trees. This is now known to be quite a mistaken idea. The ants are attracted by a viscid, sugary liquid emitted by the bugs, and which is, in fact, their excreta. This substance is being constantly shed upon the surrounding leaves, and proves very attractive, not only to ants, but to flies of all kinds, and even to bees and wasps. It is of the same nature as the 'honey-dew,' so abundantly produced by Aphides. Far from feeding upon the bugs, I believe the ants actually transport them from place to place to found fresh colonies of them in convenient situations. It is certain that the small black nest-building ant (*Cremastogaster dohrni*), that is such a nuisance on some of our estates, invariably includes in its nests colonies of Mealy-bugs (*Dactylopius*) and one or more species of *Lecanium*.

Amongst vertebrate animals birds are sometimes supposed to play an important part in the warfare against Scale Insects; but I am inclined to think that their usefulness against this particular class of insect pests has been overrated. I have watched many of our insectivorous birds in Ceylon, and I have never found them attracted by the plentiful supply of insect food spread before them in a field of 'buggy' coffee. There is a peculiar aroma about many of the Coccidæ that is possibly distasteful to birds.

Where Scale-bugs are present in large numbers, the leaves of the tree or plant will usually be disfigured by a sooty deposit, the nature of which has been frequently misunderstood by planters many of whom look upon this as the active injurious principle, instead of the mere outward indication of a less conspicuous disease. To the best of my belief, this 'black fungus' is itself absolutely innoxious to the plant. It, in fact, germinates and subsists upon the sweet liquid or honey-dew described above. In dry weather the black growth can be easily peeled off in a thin film, leaving the surface of the leaf healthy and unaltered in appearance, proving that the action of the fungus is quite superficial. The supposed injury said to be due to the choking up of the breathing pores (stomata) of the leaf, is largely illusory as the greater number of these stomata are found on their *under* surface, where the fungus never occurs, owing to the simple fact that the honey-dew can fall only upon the *upper* or exposed surface of the leaves. Nietner (*The Coffee Tree and its Enemies*, p. 8) mentions that two distinct species of the fungus have been identified from Ceylon, named respectively, *Syncladium Nietneri*, Rabenhorst, and *Triposporum gardneri*, Berkeley, both of them very similar in appearance, and both occurring on coffee in association with Scale-bug.

It is remarkable that in some species of Coccidæ, especially such as have in all probability been imported, successive generations of fertile females are produced without the intervention of any male insects. This phenomenon, known as 'Parthenogenesis,' is also found amongst Aphides, and occasionally in a modified form in higher orders of insects. During the ten or more years that the Green-bug (*Lecanium viride*) has been with us no male insects have been observed. It cannot be supposed that this state of affairs can go on indefinitely. A generation of males will probably appear in due course, when the more natural mode of propagation will probably be adopted.

CHAPTER II.

COLLECTION AND PREPARATION.

For those who may feel sufficient interest in the subject to commence a study of this group of insects, a few words as to the collection and preparation of specimens may not be out of place. With respect to the latter, what is a convenient method for one person may, of course, not prove so for another. I can only explain the plan that, after considerable experiment, I have found most convenient to myself.

For the collector the field is enormous. I think I may safely say that there is no country in the world where Coccidæ are not to be found, and scarcely a single species of plant that is not liable to their attacks.

In Ceylon and other tropical countries they are to be found at all seasons of the year. In northern countries they are naturally more abundant in the summer months, though even in winter a few species may be collected from the bare branches of trees and shrubs; in fact, some of them are more easily discovered at that time than when the plants are in full leaf. The male insects should be looked for in the early spring in such parts of the world as enjoy the normal change of seasons. In Ceylon I have usually found both sexes to occur independently of the time of year.

Any part of the plant—root, stem, branches, leaves, flower-buds, or fruit—may harbour members of the family. Most species will be found closely applied to the surface of the leaves or branches; but others conceal themselves, some in galls or irregular swellings, some imbedded in the soft tissues of the bark, others secreted in natural glandular pockets on the leaves, or enfolded within the sheathing bracts.

There are various signs that reveal to the practised eye the presence of Coccidæ. Indefinite yellowish or otherwise discoloured patches on the upper surface of a leaf will often indicate the position of some of these insects located beneath it. When the leaves of a plant are disfigured by a deposit of black fungus, we may reasonably expect to find a flourishing colony of Scale-bugs, though sometimes the same fungus follows the attacks of Aphides. By this last means alone a badly infested tree may often be noted from a long distance. Small spots of white powdery matter projecting from some crevice will often repay closer inspection, and a busy string of ants ascending the stem of a plant may point the way to an unsuspected establishment of the insects of which we are in search.

The members of this family are either without means of locomotion or slow to use their limbs when present. They are usually more or less permanently anchored to the plant by their long sucking-tubes. The leaves or stems to which they are attached should be removed with them, or pieces of the bark may be sliced off with the insects; but it is advisable not to detach the insects themselves, for, in so doing, their mouth parts are liable to be injured, and for cabinet specimens the insects must be mounted in their natural position on pieces of their food plant.

Before drying the insects and storing them away, they should be carefully examined in the living state, and notes made of their colour, markings, natural form, &c. If there are any tenanted male scales, these should be placed in small glass-topped boxes, to allow of the emergence of the winged insects. After the preliminary examination, the specimens may be killed (in a stifling-bottle) and stored away until it be convenient to examine them more thoroughly. All the minute microscopical details may be studied as well, or even better, from the dried specimens, if properly prepared in the manner described below. In the mean time the specimens may be folded—leaves and all—in some soft (preferably absorbent) paper, and placed inside a stout envelope (such as is used by florists for their packets of seeds) upon which must be noted the name of the insect (if known), the date, and the locality. The notes made at the preliminary examination may be included in the envelope. Empty cigar boxes form convenient receptacles for these packets. A few lumps of naphthaline and camphor should be placed in the box to keep out mites.

For a working collection it will be necessary to make two separate arrangements; one of dried specimens to show the external characters, and another of prepared mounts for the microscope. For the former purpose, small pieces of the leaves or branches with the attached insects may be pinned or gummed upon cards, and arranged in the drawers of a cabinet, like any other collection of insects. Such an arrangement will be found useful for rough comparison of specimens.

But for accurate determination a careful microscopical examination is absolutely necessary. There are many species that it would be quite impossible to separate by their external characters alone. A rather more elaborate preparation is necessary to exhibit the structural details. The specimen to be examined has first to be boiled for a few minutes in some strong alkali, such as liquor potassæ, over a spirit-lamp, the time to depend upon the size of the object. As soon as the specimen begins to look clear and skeletonised, the boiling should cease. The process will be assisted by previously making a small opening in the body of the insect, to allow of the soft parts being expelled. Large species such as *Walkeriana, Icerya*, &c., can be boiled in a wide test-tube or beaker; for smaller insects I find it more convenient to use a deep watch-glass, from which they can afterwards be more easily removed. This boiling process will also serve to restore the shrivelled specimen to its normal form and dimensions. From the potash they should be shifted to a vessel (another watch-glass) containing distilled water, the specimens being lifted with the flattened point of a piece of thin wire (a feather or brush would be rapidly destroyed by the caustic action of the potash). Here they may remain for an hour or more, by which time the potash will have been removed together with more of the decomposed contents of the body. Next lift the insect carefully on to a glass slide, depositing it in a drop of alcohol (or rectified spirits of wine), arrange the parts conveniently for examination, using a fine camel's-hair brush or snipe's feather; put a drop of weak glycerine on a thin cover-glass, and lower this gently on to the object, which is now ready for a further examination. I find glycerine more suitable than alcohol, as the latter evaporates so rapidly, while the former will remain for days or weeks without appreciable reduction in quantity, so that specimens may be kept under observation for some time without further treatment. With this medium all the minute details of structure can usually be distinguished, and carefully enlarged drawings of the limbs, spinnerets, marginal processes, &c., should now be made. I find the $\frac{2}{3}$, $\frac{1}{4}$, and $\frac{1}{6}$ inch micro-objectives the most convenient for the purpose. Higher powers are seldom required. The more delicate hairs and spinnerets are sometimes so very transparent that staining is necessary to make them more visible. For this purpose an alcoholic solution of magenta, or some other aniline dye (or even ordinary red ink), may be run in under the cover-glass, and the slide put aside for twenty-four hours. Finally, the object should be washed in clean alcohol to remove the superfluous colour, and permanently mounted for microscopic use. I would strongly recommend the staining of all the more delicate objects. Unsuspected characters will often be revealed by this means, and it will help to clear up many doubtful points. If there is any question about the number of joints in an antenna, or the hairs on the anal ring, the addition of the stain will usually make them perfectly distinct.

Canada balsam is usually advocated for mounting prepared specimens of Coccidæ. I find, however, that this medium has the property of making delicate objects so very transparent that much of the minute detail is lost or extremely difficult to decipher unless they have previously been deeply stained. Balsam has also the disadvantage of darkening with age. Glycerine jelly or Farrant's Medium—personally I prefer the latter—have not these faults; and the object may be transferred direct from the alcohol to either of these mixtures. Care should be taken that no more of the medium is used than will suffice to fill the very small space between the cover-glass and the slide; otherwise it will be found difficult to seal up the mount satisfactorily. If the Canada balsam method is preferred, the specimen must be transferred from the alcohol to oil of cloves or cedar-wood oil, to remove all traces of the spirit; after which it may be mounted in balsam in the usual way. Whichever preparation is adopted, the best results will be obtained by placing a drop of the medium on the cover-glass and lowering it on to the object. After placing the cover-glass in position the medium can be induced to spread evenly, and any air-bubbles can be expelled by gently heating the slide over a spirit-lamp. Canada balsam will set quite hard and requires no final sealing; but with the glycerine preparations a ring of asphaltum or caoutchouc cement must be formed round the edges of the cover-glass, to prevent subsequent evaporation or exudation of the medium. No doubt, in some particulars, Canada balsam has its advantages. Such mounts are more permanent and less liable to injury; and, where the object is at all opaque, the greater refractiveness of the balsam helps to clear it. A duplicate set of specimens might with advantage be mounted in each medium.

The winged males of nearly all Coccidæ are very minute and fragile. They can be satisfactorily examined only in the fresh state. They live but a few hours and shrivel rapidly after death. Any notes and drawings should therefore be made as soon after the emergence of the perfect insect as possible. They do not respond to the treatment recommended for the female insects, and permanent mounts either in balsam or glycerine medium are seldom satisfactory.

These few hints may possibly be of assistance to the beginner (I do not presume to offer them to more experienced workers), but the particular method that will produce the best results in his case can be learned only by personal experience, and each individual will doubtless find out for himself many little devices to assist him in his work.

EXPLANATION OF PLATE I.

STRUCTURAL CHARACTERS OF FEMALE AND LARVAL DIASPINÆ.

Fig. 1. *Chionaspis*, ♂ and ♀ scales.
    2. *Aonidia*,   ♂ ,, ♀ ,,
    3. *Fiorinia*,   ♂ ,, ♀ ,,
    4. *Diaspis*,   ♂ ,, ♀ ,,
    5. *Mytilaspis*, ♂ ,, ♀ ,,
    6. *Parlatoria*, ♂ ,, ♀ ,,
    7. *Aspidiotus*, ♂ ,, ♀ ,,
    8. 1st pellicle of *Aspidiotus*, from below.
    9.   ,,   *Diaspis*,   ,,   ,,
  10. *Diaspis rosæ*, ♀, ventral view.
        (A) head, (B) prothorax, (C) mesothorax, (D) metathorax.
        (I., II., III.) abdominal segments, (IV.) 'pygidium.'
        (*a*) antennæ, (*b*) rostrum, (*c*) mentum, (*d*) anterior spiracles, (*e*) rostral setæ, (*f*) 2nd pair of spiracles, (*g*) oval pores of tubular spinnerets, (*h*) circumgenital glands.
  11. Antennæ of adult ♀ *Diaspinæ*.
        (*a*) *Fiorinia fiorinæ*, (*b*) *Aspidiotus osbeckiæ*, (*c*) *Aspidiotus limonii*, (*d*) *Chionaspis aspidistræ*.
  12. Egg of *Diaspis*.
  13. Antennæ of larval *Diaspinæ*.
        (*a*) *Parlatoria zizyphi*, (*b*) *Diaspis boisduvallii*, (*c*) *Mytilaspis pomorum*, (*d*) *Aspidiotus ficus*.
  14. Diagram of pygidium (combining characters of various genera of Diaspinæ).
        (*a*) Circumgenital glands, median group, (*b*) upper lateral groups, (*c*) lower lateral groups, (*d*) genital orifice, (*e*) anal orifice, (*f*) tubular spinnerets, trumpet-shaped, (*g*) filiform spinnerets, (*h*) cylindrical spinnerets, (*i*) lobes, (*j*) marginal prominences, (*k*) squames, fimbriated, (*l*) spine-like squames, (*m*) spines.
  15. Tubular spinneret, with glands. (Adapted from drawings by Berlese.)
        (*a*) Silk gland, (*b*) accessory glands (said to secrete varnishing or cementing material), (*c*) capitate head of chitinous tube, (*d*) chitinous tongue, communicating with silk duct, (*e*) chitinous tube, connecting silk duct with the surface, (*f*) external orifice of spinneret.
  16. Squames : (*a*) spiniform, (*b*) spiniform, with divided extremity, (*c*) with narrowly fimbriated extremity, (*d*) with broadly fimbriated extremity, (*e*, *f*) laterally fimbriate.

CHAPTER III.

CHARACTERS AND CLASSIFICATION.

THE Coccidæ—commonly known by the name of Scale Insects or Scale-bugs — are a subdivision of the order Hemiptera. This order comprises two sub-orders, the Homoptera and Heteroptera, both of which are characterised by the presence of suctorial mouths and an incomplete metamorphosis.

The Coccidæ are considered to belong to the first of these divisions—the Homoptera. There is, however, one point in which they do not conform with that group. The wings of typical Homoptera are not carried horizontally, but slope upwards and inwards, their inner margins meeting above the middle of the back without overlapping. We can only judge of this character in the Coccidæ by the winged males, and in these the wings are carried horizontally above the back of the insect when at rest, often completely overlapping each other—a character according rather with the Heteroptera.

The Coccidæ may be distinguished from their nearest allies by the following characters :—

1. The absence of wings in the females (*pl.* I. *fig.* 10).
2. The presence in the adult males of only two wings, supplemented by a pair of hooked organs representing the missing hind wings (*pl.* II. *fig.* 2); which organs are homologous with the 'halteres' of Diptera, but, unlike the latter, are connected with the wings and serve to strengthen them during flight.
3. The legs in both sexes (when present) terminating in a single claw and having normally only one joint in the tarsus (*pl.* II. *fig.* 6). (An exception occurs in the abnormal genus *Exæretopus*, Newst., in which the anterior tarsi are two-jointed; and Professor Cockerell has quite recently described a gall-making Coccid from America, *Oliffiella cristicola*, in which the same peculiarity exists.)

4. The absence of any mouth or feeding apparatus in the adult males, which are instead usually supplied with large supplementary eyes (*pl.* II. *fig.* 5).

There are insects belonging to the *Psyllidæ* and *Aleurodidæ* that very closely resemble some Coccidæ in outward appearance; but an examination of the above characters will at once enable an observer to separate them, the members of both those families possessing two claws on the feet, and having four wings in the adult stages of both sexes. The single-clawed foot of the Coccidæ, though not quite a unique character among insects, will be found almost sufficient in itself to distinguish a Coccid from any other insect that might otherwise be mistaken for one of this family.

Other general characters belonging to the family are:—

5. The absence of any definite boundary between the head and thorax in the female (*pl.* I. *fig.* 10).
6. The rostrum, or mouth-parts, situate far back on the under surface of the insect between—or even behind—the insertion of the first pair of legs.
7. The secretion in greater or less quantity of waxy, fibrous, or resinous matter.

The life-history and metamorphosis of the Coccidæ vary somewhat in the several sub-families and, together with the more specialised characters, may be more conveniently described in the chapters dealing separately with those sub-families.

The family Coccidæ has been variously divided into several sub-families varying in number from the four of Signoret— *Diaspidæ, Brachyscelidæ, Lecaniidæ,* and *Coccidæ*—to the ten divisions in Professor Cockerell's latest Check List,' viz., *Monophlebinæ, Porphyrophorinæ, Coccinæ, Hemicoccinæ, Ortheziinæ, Asterolecaniinæ, Brachyscelinæ, Idiococcinæ, Lecaniinæ,* and *Diaspinæ.* Of this larger number I should be inclined to include the *Porphyrophorinæ* with the *Monophebinæ.* Judging from Signoret's description, the characters of both male and female of the genus *Porphyrophora* point to its close alliance with *Cœlostoma* of Maskell. Nor does the separation of the *Asterolecaniinæ* from the *Coccinæ* (as generally constituted) seem necessary. The principal difference appears to lie in the nature of the secreted test, which in *Asterolecanium* or *Planchonia* is of a transparent horny texture through which the body of the insect can be plainly discerned, while in the majority of the Coccinæ the sacs are formed of an opaque felted or woolly secretion.

This sub-family, *Coccinæ*, presents considerable difficulties to the systematist if made to include such divergent forms as typical *Coccus* on the one hand and *Eriococcus*, or *Dactylopius*, on the other. The genus *Coccus*, which must naturally be taken as the type of the sub-family, contains insects possessing neither the anal tubercles nor the setiferous anal ring of the other genera usually associated with it. It might with more justice form the type of a group including *Gymnococcus* Dougl., *Cryptococcus* Dougl., *Capulinia* Sign., and perhaps *Xylococcus* Low. Mr. Maskell's division of *Idiococcinæ* is with difficulty separable from such a group, but *Sphærococcus* and *Cylindrococcus* contain such abnormal and extraordinary forms that they may conveniently be retained in a separate division. The Dactylopiid group (including *Planchonia*), characterised by the setiferous anal ring and tubercles, form a natural division which, in the classification adopted below, is called *Dactylopiinæ*.

Conchaspis and *Tachardia*, two genera at present included in the *Coccinæ* and *Brachyscelinæ* respectively, bear such distinct characters in all their stages as to warrant their accommodation in separate sub-orders.

The following classification is put forward tentatively. I am fully aware of its imperfections; but by this arrangement the several sub-families not only fall into more or less naturally allied groups, but lend themselves to a convenient system of Synopsis.

Two primary divisions may be made, according to the presence or absence of compound (facetted) eyes in the adult males. In the majority of the sub-families the male insect possesses four simple eyes, two on the upper and two on the under surface of the head. These are really supplementary eyes, homologous with the ocelli of other insects, the true eyes in these cases being very much reduced, appearing merely as a small colourless tubercle on each side of the head, or entirely obsolete. In the division containing males with compound visual organs, these represent the true eyes occupying their normal position on the sides of the head, while the ocelli are small or altogether wanting.

SYNOPSIS OF SUB-FAMILIES.

A. Males with simple eyes.
    a. Abdomen of female terminating in a compound segment (pygidium). Anal orifice without a setiferous ring.
        (i.) Insects with a separate covering-scale formed entirely of secretionary matter without admixture of the exuviæ.

Adult female retaining limbs and antennæ. Mentum dimerous .... CONCHASPINÆ

(ii.) Insects with a separate covering-scale composed partly of the exuviæ and partly of secretionary matter. Adult female without limbs. Mentum monomerous. DIASPINÆ.

β. Abdomen of female without a definite pygidium. Anal orifice with a definite setiferous ring.

(iii.) Females with posterior extremity cleft. Anal orifice closed above by a pair of triangular hinged plates forming a valve .... LECANIINÆ.

(iv.) Adult females with cleft extremity and anal plates as in *Lecanium*. Larvæ with abdominal lobes as in *Dactylopiinæ* .... HEMICOCCINÆ.

(v.) Abdominal extremity not cleft; usually with a pair of more or less prominent rounded tubercles, each bearing a long seta. No hinged plates above anal orifice.

DACTYLOPIINÆ.

γ. Insects enclosed in a resinous cell with three orifices. Adult female apodous, with the terminal segments produced into a tail-like organ bearing at the extremity the anal orifice, which is surrounded by a broken setiferous ring. A prominent spine-like organ above the base of the caudal extension.

(vi.) .... TACHARDIINÆ.

δ. Females without anal tubercles. No setiferous anal ring.

(vii.) .... COCCINÆ.

ε. 'Adult females active or stationary; gall making, or naked, or producing cotton or wax. Anal tubercles entirely absent; anal ring hairless. Antennæ with usually less than seven joints. Body not prolonged posteriorly.' (Maskell, *Trans. New Zealand Inst.*, 1892, p. 236). Larvæ with anal tubercles, adult without.

(viii.) .... IDIOCOCCINÆ.

ζ. Insects enclosed in galls. Limbs either persisting, rudimentary, or obsolete.

(ix.) .... BRACHYSCELINÆ.

B. Males with compound eyes.

α. Females with definite setiferous anal ring.

(x.) .... ORTHEZIINÆ.

β. Females without definite setiferous anal ring.

(xi.) .... MONOPHLEBINÆ.

D

EXPLANATION OF PLATE II.

Structural Characters of Male Diaspinæ.

Fig. 1. Pupa, ventral view.
    2. Adult ♂, dorsal view.
        (*a*) Head, (*b*) prothorax, (*c*) mesothorax, (*d*) metathorax, (*e*) abdomen, (*f*) dorsal scute of mesothorax, (*g*) apodema, (*h*) scutellum of metathorax.
    3. 'Halter,' dorsal view.
        (*a*) Basal portion, (*b*) hooked bristle.
    4. Head of adult ♂, dorsal view.
        (*a*) Upper ocelli, (*b*) rudimentary eyes, (*c*) genæ.
    5. Head, from below.
        (*a*) Lower ocelli, (*b*) rudimentary eyes, (*c*) genæ.
    6. Leg of adult ♂.
        (*a*) Coxa, (*b*) trochanter, (*c*) femur, (*d*) tibia, (*e*) tarsus, (*f*) claw, (*g*) ungual digitules, (*h*) tarsal digitules.
    7. Adult ♂, ventral view.
        (*a*) Head, (*b*) prothorax, (*c*) mesothorax, (*d*) metathorax, (*e*) abdomen, (*f*) divided scute of mesothorax.
    8. Adult ♂, side view.
        (*a*) Head, (*b*) prothorax, (*c*) mesothorax, (*d*) metathorax, (*e*) abdomen.
    9. Antenna of adult ♂.
   10. Genital sheath of adult ♂, from below, showing elongate valve.

CHAPTER IV.

Conchaspinæ.

As briefly diagnosed in the Synopsis at the end of the previous chapter, the sub-family *Conchaspinæ* includes insects having a separate covering-scale, formed entirely of secretionary matter without admixture of the exuviæ, the adult female retaining limbs and antennæ. The mentum dimerous.

This division is represented by a single genus only — *Conchaspis*, the characters of which (as far as they are known) are fully described below.

Conchaspis, *Cockerell.*

Conchaspis, Cockerell.—'Coccidæ, or Scale Insects,' *Bulletin Bot. Depart. Jamaica*, Feb. 1893, p. 9.

Pseudiglisia, Newstead.—'Notes on New or Little-known Coccidæ,' *Ent. Monthly Mag.*, July, 1893, p. 158.

This genus is so aberrant, and so few species are known, that no very definite diagnosis can at present be drawn up. Mr. Cockerell has, apparently, given no formal description of the genus. Mr. Newstead gave a short diagnosis in founding his genus *Pseudiglisia*. His description runs as follows:—'Scale elevated, more or less circular, ridged; ventral scale complete, detached; antennæ of four joints; anal lobes very minute; last five segments of body with broad, chitinous plates, bearing spinnerets; rostrum biarticulate.' Even this broad definition is too narrow to admit the undoubtedly congeneric insect described below, for the Ceylon representative bears three-jointed antennæ, and its scale is not ridged.

I would suggest the following as a temporary diagnosis:—Scale elevated, more or less circular; adult female retaining limbs and antennæ, the latter of few joints; genital aperture without setiferous ring; mentum biarticulate; terminal segments of body united into a piece somewhat resembling the pygidium of the *Diaspinæ*.

In the three species at present recognised there appears to be no tibiotarsal articulation, and I am inclined to think that this character will prove of generic importance. It is true that Mr. Newstead, in his description of *C. rodriguesiæ*, says:—'Tibiæ grooved, about as long again as the tarsi;' but that author informs me that the supposed articulation proves to be little more than a marginal indentation. This fused condition of the tibia and tarsus is very marked in both sexes of *C. socialis* from Ceylon; it also holds good with *C. angraci*, Ckll. from America.

The male is known only from the Ceylon species. The pupa is contained in an oblong felted sac of firm consistency. The adult male has four ocelli, the antennæ bear several knobbed hairs at the apex, the genital spike is long, slender, and pointed. The tibia and tarsus are fused together as in the female.

The larvæ (as known from *C. rodriguesiæ*) appear to differ but slightly from the adult female, the principal difference being in the greater number of joints in the antennæ. The eggs are proportionately large.

It is difficult to assign this genus to any of the recognised sub-families. More material is wanted, and other species must be studied before a decision would be warranted. It seems to be widely separated from any of the known genera. The scale, though superficially Diaspid in appearance, is without the cast larval skins (pellicles) that enter into the construction of the puparia of all the members of that sub-family. The two-jointed mentum points to some affinity with the *Coccinæ*, and there this genus may perhaps eventually find a resting-place; but, pending further light on the subject, the genus is here provisionally isolated by placing it at the commencement of the list, before the *Diaspinæ*.

CONCHASPIS SOCIALIS, *sp. nov.*

(Plates III. and IIIA.)

Female puparium comparatively large; circular; rounded or bluntly conical above (*fig.* 10); firm, opaque, and closely felted; outer surface greyish or greyish brown (*fig.* 8); inner surface white and smooth (*fig.* 9). The ventral scale is represented merely by a thin whitish film adhering to the plant (*fig.* 9). Diameter of mature scale about 5 mm.

Male puparium (*fig.* 7, *a*) white, oblong, flattened, closely felted, completely enveloping the pupa. The hinder extremity has a valvular opening. Length 1·50 mm. Breadth, at widest part, about 0·75 mm. The male puparia do not occur separately, but are always collected together, in groups of ten or more, beneath the parent scale (*fig.* 7) which they completely fill.

Adult female reddish brown (*fig.* 9). Surface chitinous and shining, minutely rugose with lines of irregular raised spots. Form oblong (*figs.* 11, 12), rounded in front, abruptly narrowed behind the thorax; segments well defined. Eyes (*fig.* 13) large, situated dorsally each in a conspicuous circumscribed oval space of a paler and clearer colour than the surrounding area. Body terminating in a blunt point; the extremity fringed dorsally with eight short pointed processes (*fig.* 18), the margin thickened and strongly chitinised (*fig.* 19). On the lateral margins of each of the three last segments is a small broadly rounded chitinous transparent lobe (*fig.* 19). Each segment of the body bears a lateral tuft or group of three or four stout hairs, those on the anterior parts longest, the others diminishing as they approach the extremity. The abdominal segments are strengthened by broad bands of firmer texture, especially towards the extremity, where they are interrupted in the centre, and take the form of distinct circumscribed plates. The last dorsal pair of these plates simulate the anal lobes of the Lecaniinæ, but they are continuous with the general surface of the tegument, and have no free extremity. Antennæ situated within the margin on the ventral surface (*fig.* 12), consisting of three distinct joints only (*fig.* 20), though there are sometimes traces of an indistinct subdivision of either the terminal or second joint: first joint short, constricted at the base; second joint longest; extremity of third joint truncate, bearing a crown of short stout hairs, of which one is considerably thicker than the others; the first and second joints have each a single longish stout hair on the side. Rostrum large and conspicuous; mentum biarticulate.* Legs stout and well developed (*fig.* 14), extending considerably beyond the margin

---

* By a slight error of the lithographic artist the mentum (*in fig.* 12, *pl.* IIIA) is made to appear triarticulate. An extra joint has also been added to the antennæ in the same figure, making them 4- instead of 3-jointed.

of the body; no tibio-tarsal articulation (*fig.* 15); femur much longer than tibio-tarsus and claw together; one or two long fine hairs on each joint; claw short and stout, without digitules, but inner margin dilated at base. There are apparently three pairs of spiracles, one situated at the base of each leg; the orifices of the first two pairs are accompanied by three large glandular pores; the spiracles on the metathorax are simpler in structure, being mere cup-shaped depressions, and are possibly not functional. There is a ventral group of five or six compound glandular pores (*figs.* 12 and 19) towards the lateral margin of each of the first three abdominal segments, and small groups of short broad compressed tubular ducts (*figs.* 16, 17) opening dorsally on the margin of the first four abdominal segments. The genital aperture opens ventrally on the posterior edge of the fifth abdominal segment, and is covered by a semicircular lobe (*fig.* 19). I have been unable to determine the position of the anal aperture. Length, 1·25 mm.; breadth, 0·80 mm. The small size of the adult insect as compared with the proportionately large scale is remarkable. This roomy covering appears to be provided for the protection of the male pupæ.

Adult male (*fig.* 1) in form not unlike the males of *Planchonia*. Colour: head and thorax pale reddish yellow, abdomen yellowish white. Head (*fig.* 3) rounded in front, widest behind the ocelli. Ocelli four, rather small. Antennæ (*fig.* 4) with seven joints; first two joints small; second with two stoutish hairs on one side; third longest; fourth to sixth with four knobbed hairs at apex, which is rather truncate; all the joints with numerous short bristles or spicules. Wings long and ample. Genital spike very long and slender, rather more than half the length of the body. Legs moderately long; femur rather short; tibio-tarsus long and slender without any trace of division (*fig.* 2); foot with four digitules, all slender knobbed hairs, those on tarsus longest; claw rather slender. Length 1·25 mm.

I have unfortunately been unable to obtain young larvæ or the earlier stages of the female. Some of the male sacs contained male larvæ which were oval in form (*fig.* 5), with four-jointed antennæ (*fig.* 6).

Comparatively large purplish eggs were present under such of the scales as were not choked up with the male puparia. And there was nearly always present a loose yellowish pellicle (*fig.* 9), which I at first supposed to be the remains of an earlier stage of the female insect. But a more careful examination showed it to be of hymenopterous origin—presumably the empty pupa of some parasite. It is curious, however, that the female Coccid seemed in every case to be uninjured.

My examples of this interesting species were collected by Mr. John Pole on an unidentified shrub at Tangalla. The scales look like warty excrescences on the twigs of the shrub, and in spite of their size are very inconspicuous.

The specific name is suggested by the social habit of the male larvæ and pupa, which, together with one or more females of the same generation, are grouped beneath the covering scale of the parent insect.

EXPLANATION OF PLATES III. AND IIIA.

CONCHASPIS SOCIALIS.

(All figures, except No. 10, more or less enlarged.)

Fig. 1. Adult male, side view.
2. ,, ,, foot.
3. ,, ,, head, from above.
4. ,, ,, antenna.
5. Male larva.
6. ,, ,, antenna.
7. Female scale from below, showing group of male puparia, and a young female.
7A. A single male puparium, detached from the mass, upper side
8. Female scale, from above.
9. ,, ,, turned back, showing adult female and eggs *in situ*.
10. Twig, with female scales, natural size.
11. Adult female, dorsal view.
12. ,, ,, ventral view.
13. ,, ,, head, from above, showing eyes.
14. ,, ,, leg.
15. ,, ,, foot.
16. ,, ,, margin of abdominal segment, showing tubular spinnerets.
17. ,, ,, a single spinneret, seen from the side.
18. ,, ,, extremity of body, dorsal view.
19. ,, ,, terminal segments, ventral view.
20. ,, ,, antenna.

CHAPTER V.

DIASPINÆ.

'INSECTS having a separate covering-scale composed partly of the exuviæ and partly of secretionary matter. Adult female without limbs. Mentum monomerous.'

The form of the scale—or puparium, as it is often called—varies from circular to oblong, or even linear. Its structure is in all cases practically identical, and can be best understood by following the life-history of a typical Diaspid from the egg to the adult stage.

The eggs, in the oviparous species, are deposited beneath the parent scale, and are retained there until hatched. Some few species are ovoviviparous, and in such cases the young make their escape at once from beneath the edges of the scale.

The small oval larvæ are at first very active and of a wandering disposition, with well-developed limbs and antennæ. They at once set out in search of a suitable spot upon which to fix themselves for life; for, having once settled down, they cannot afterwards renew their wanderings, but remain attached to the same spot by their sucking-tubes, which are buried deep in the tissues of the plant. Soon after the insect has become stationary, a very thin layer of waxy secretion is deposited over the dorsal surface, sometimes forming a thickened central boss or a raised ring.

The first larval skin or 'pellicle' is shed comparatively early in the life of the insect. A curious difference in the mode of effecting the ecdysis is noticeable between *Aspidiotus* and other genera. In *Aspidiotus* the larval skin splits along the margins, completely separating the dorsal and ventral halves, which subsequently become incorporated respectively into the dorsal and ventral scales of the puparium. In this case the visible part of the pellicle will be destitute of the sheaths of the antennæ or limbs (*pl.* I. *fig.* 8). In all other genera that I have observed the whole of the larval skin is attached to the dorsal scale of the puparium, the rupture occurring on the under surface of the head at a point between the antennæ and the rostrum, the antennæ remaining attached to the anterior margin, while the rest of the ventral parts, with the limbs and rostral apparatus, are pushed back to the posterior extremity (*pl.* I. *fig.* 9). This character will be found of assistance in determining some doubtful cases. When the male puparium is unknown, it is sometimes difficult to decide whether a particular species should be assigned to the genus *Aspidiotus* or *Diaspis*. If upon examination the larval pellicle is found to consist of the dorsal parts only and no antennæ are present, the insect may with confidence be placed under *Aspidiotus*. But, if, on the contrary, the antennæ are found to be still attached to the anterior margin of the pellicle, the insect may be considered a *Diaspis*. In all cases the harder dorsal parts retain their original form and position.

So far the puparia of both sexes have undergone a similar development; but from this point they diverge. With the first moult the insect has discarded its limbs and assumed a pupiform state, which in the case of the female is retained, with some modification, during the rest of its life. It will be convenient to follow the development of the female puparium first.

After shedding the first skin the insect rapidly increases in bulk, so that its soft body can no longer be protected or concealed by the larval scale. To supply the deficiency, it extends the margin of the discarded pellicle by the addition of a thin membrane-like material composed of closely woven filaments secreted by special organs (to be described later) situated on the terminal segments of the body. This thin covering-scale keeps pace with the growth of the insect. At the approach of the second and last moulting period, the dorsal parts of the body become more rigid and horny, the skin splits as before, the pellicle remaining in position attached to the dorsal scale, and the insect assumes its final stage, which does not differ very greatly in external characters from the preceding one.

There are not the same differences in the mode of shedding the second pellicle that were noticeable at the earlier ecdysis. In all the genera the rupture occurs on the under surface in front of the rostrum, and the ventral skin is pushed back to the posterior extremity, where it remains attached to the hardened dorsal parts.

The adult female is at first concealed by the two discarded pellicles, which are cemented together by the thin membranous coating secreted during the previous stage. It is at this time, before the further enlargement of the scale, that impregnation takes place. The fecundated female soon outgrows its former covering, which is accordingly supplemented by the further formation of the membrane-like secretion mentioned above. This supplementary part may be of stouter or finer texture, opaque or semi-transparent, colourless or coloured, according to the species; but it is produced in practically the same way from special tubular organs or spinnerets, situated chiefly on the margins and dorsal surface of the terminal segment or pygidium. In this later stage the extension of the puparium is continued until it is of sufficient size not only to protect the soft-bodied insect itself, but to form a receptacle for the numerous eggs that are now deposited there. The so-called ventral scale is really continuous with the upper part, and forms a bed upon which the insect rests; but this part is usually very much thinner, often consisting merely of a delicate film adhering closely to the surface of the plant, though occasionally it is of almost as firm a texture as the upper part.

In some genera, such as *Aspidiotus* and *Diaspis*, the secretionary supplement completely surrounds the pellicles (*pl.* I. *figs.* 7, 4); in others, such as *Mytilaspis* or *Chionaspis*, it is extended in a backward direction only (*pl.* I. *figs.* 5, 1). In the former case, the insect must revolve completely around the point of attachment during the construction of the puparium. In the genera forming elongate puparia, a to-and-fro sweeping motion of the hinder parts of the body is sufficient to produce the resulting form of scale.

The abnormal genera *Aonidia* and *Fiorinia* have a slightly different development. In these the insect attains its greatest dimensions during the second stage. At the time of the second moult the skin is not actually shed, nor is it even ruptured, but completely encloses the body of the adult female, which becomes gradually reduced in size and shrinks away from its former skin, the vacant space, in the case of *Fiorinia*, being subsequently packed with the ova. It is remarkable that, although the skin of the enlarged second pellicle is entire in these two genera, the rostral apparatus is still displaced to the posterior extremity, indicating a most extraordinary distension of the skin of the anterior parts.

In some species the puparia have the appearance of being situate beneath the cuticle of the plant; but this is seldom really the case, the effect being produced by the presence in the scale of hairs or loose fibrous matter superficial to the actual cuticle.

The flattened trowel-shaped pygidium of the insect enables it to force its way beneath the hairs and fibres, gradually raising and incorporating them into its scale without disturbing their original position. The puparia of *Chionaspis biclavis* and *Chionaspis elæagnus* are excellent examples of such a habit.

The male puparium in such genera as *Aspidiotus* and *Mytilaspis* is constructed on the same plan as that of the female, the secretionary supplement being of nearly the same colour, texture, and form in both sexes (*pl.* I. *figs.* 7, 5). The process, however, stops at an earlier stage, only one (the first larval) pellicle being utilised in the formation of the complete scale.

In some genera—such as *Diaspis*, *Chionaspis*, and *Fiorinia*— there is a more marked contrast between the scales of the two sexes. In these genera the male puparium is always more linear in form, with sub-parallel sides (*pl.* I. *figs.* 4, 1, 3); the solitary pellicles occupy the anterior extremity; the supplementary area composed of an opaque snowy-white secretion of a looser texture, and frequently ornamented with more or less prominent thickened carinæ of the same substance.

Although only one pellicle appears on the male puparium, there are altogether three changes of skin before the emergence of the adult male. The first pellicle is shed in the same manner as that of the female, the same generic differences being noticeable in the line of rupture. The pellicles of the two latter moults are discarded from the hinder extremity of the puparium.

The fact that the male puparia are often clustered together in large numbers by themselves points to the probability of their being in some cases separate broods consisting entirely of males.

It will now be convenient to examine the structure of the insect itself in more detail, returning once more to the female side and studying it from the earliest stages.

The egg (*pl.* I. *fig.* 12) is always of a more or less oblong oval form, its surface usually dusted with minute waxy granules.

The young larva is oval and flattish. The segmentation is not very decided, but the usual divisions of the thorax can be made out and six segments can be distinguished in the abdomen, the sixth probably consisting of several segments joined together. Even at this early stage the terminal segment is fringed with small lobes and squames, though not to the extent that is found in the pygidium of the adult female. Dr. Antonio Berlese states that, 'though these appendages will not serve to properly distinguish one species from another, they are, nevertheless, constant in their characters, and differ in the larvæ of different genera.' A pair of stout setæ, sometimes more than half as long as the body of the insect, spring from the posterior extremity. The anal orifice is situated on the dorsal surface of the terminal segment.

The antennæ spring from the ventral surface close to the anterior margin. Dr. Berlese has pointed out that, while the larvæ of some Diaspids have five-jointed antennæ (*pl.* I. *fig.* 13, *a*, *d*), in others six free joints are distinguishable (*fig.* 13, *b*, *c*). The first four or five joints are short; but the terminal one, which Dr. Berlese calls 'the funicle,' is usually as long as or longer than all the others combined, and very much wrinkled transversely. Mr. Maskell considers that six is the normal number of antennal joints in all larval Coccids. When a smaller number occurs, it is, doubtless, due to the coalition of two or more joints. In the *Diaspinæ* the missing joint must be looked for in the funicle. Each of the basal joints bears one or two stout hairs, and from six to eight hairs spring from the long terminal joint.

The eyes are minute, but usually distinct. In the larval pellicle the corneæ can always be distinguished on the anterior margin.

The comparatively large rostral apparatus is situated between the coxæ of the anterior pair of legs. The very long sucking-tube is either exserted, or withdrawn and coiled in a loop which sometimes extends far down into the abdomen of the insect.

The legs are attached to the ventral surface at some distance from the margin, but are sufficiently long to extend well beyond the edges of the body. The leg consists of the usual parts,— coxa, trochanter, femur, tibia, single-jointed tarsus, and single claw. There are four longish knobbed hairs on the foot, two springing from the base of the claw on the inner side, and two from the end of the tarsus on the outer side.

The female of the second stage differs but slightly from the adult insect. The legs have entirely disappeared, and the antennæ are reduced to minute tubercles bearing one or two stout bristles. The pygidium is well developed and provided with lobes, squames, and tubular spinnerets as in the adult; but there are no grouped glands nor any external genital orifice. The body is at first altogether soft and flexible; but the dorsal parts become hardened at the time of the second moult. Normally the insect does not increase very greatly in size during this stage, the second pellicle having usually not more than twice the diameter of the first. But in the genera *Aonidia* and *Fiorinia* it is greatly enlarged, the female insect attaining its greatest dimensions during this second stage (*pl.* I. *figs.* 2, 3).

The adult female varies considerably in form in different genera and even species. The body is either discoidal as in many species of *Aspidiotus*, more or less oblong as in *Chionaspis*, or linear as in *Ischnaspis* and some forms of *Mytilaspis*. The skin usually remains soft and flexible throughout the adult stage, though in some few species it becomes highly chitinised and indurated.

It is probable that this final stage really represents the pupa, the further development of the female insect being suppressed, although the reproductive system is complete. It is certain that this is the stage corresponding with the true pupa of the male insect.

As in the previous stage, there is no vestige of the limbs and the antennæ remain only in the form of minute setiferous tubercles (*pl.* I. *fig.* 11).

The spiracles are in two pairs, opening on the under surface. They have usually an intersegmental position, between the head and prothorax and the mesothorax and metathorax respectively, though they probably belong to the hinder segment in each case. The anterior pair is frequently situated close to the rostrum; the openings of the second pair are usually more widely separate. The anterior stigmatic orifices are often — and more rarely the posterior pair also — associated with a small group of glandular pores known as the parastigmatic glands.

The rostral apparatus, which is ventrally situate at a point near the middle of the cephalothorax, consists of a complicated chitinous piece forming the rostrum, which probably represents a fused clypeus and labrum (*fig.* 10, *b*), and a small conical mentum (*fig.* 10, *c*), which in this family is invariably of a single joint (monomerous). The four long curling setæ (*fig.* 10, *e*), which together form the sucking-tube, arise from the rostrum and pass along a channel on the upper surface of the mentum, emerging at a point near its extremity. The rostral setæ are considered to represent the maxillæ and mandibles of the insect. They can be withdrawn into the body, when they lie in a loop which extends into the abdominal parts of the insect. In prepared specimens two of the setæ usually become separate, while the other two remain in close contact. For a more detailed description of the parts of the rostrum Dr. Berlese's excellent work should be consulted.

The divisions of the head and thorax are ill defined. The abdomen consists of a certain number of free basal segments and a large compound flattened terminal piece known as the pygidium, which is usually triangular in form and more highly chitinised than the rest of the body. Taking fourteen as the typical number of somites in an insect, of which the first four go to make up the head and thorax, we have ten segments allotted to the abdomen. But sometimes only two free abdominal segments can be distinguished above the pygidium of the Diaspinæ, and never more than six. We have, then, from four to eight segments to be accounted for in the space occupied by the pygidium. It is difficult to make out the full number of divisions, and probably some of the segments have been reduced to infinitesimal proportions. In some species there are regular series of oval pores and interrupted transverse lines indicating the boundaries of suppressed segments, and it is probable that the various marginal lobes and groups of spine-like processes are each referable to one of these divisions. It is, however, as Dr. David Sharp points out (*Cambridge Nat. Hist. Insects*, Part I. p. 89), at present premature to say that all insects are made up of the same number of primary segments.

It is in this third and final stage that we find for the first time an external genital orifice, which is usually situated about the middle of the under surface of the pygidium, though in the genus *Aonidia* it will be found nearer the base of the segment. Around the genital orifice are often disposed several groups of glandular organs whose function is still rather obscure. They have been variously termed by different authors, 'grouped spinnerets,' 'grouped abdominal glands,' 'wax glands,' and 'circumgenital glands,' of which the last term is perhaps the most correct. There are usually five distinct groups (*pl.* I. *fig.* 14), distinguished as the median (a), the upper laterals (b), and the lower laterals (c). In some species of *Fiorinia* the median and upper laterals are united into a continuous arch. In many species of *Aspidiotus* there are four groups only, the median being absent. In *Poliaspis* two concentric series of groups are present, and there are examples in nearly every genus in which the circumgenital glands are entirely wanting. Their absence in many species proves that they are not concerned in the secretion of the puparium. They are situate inside the body towards the ventral side, and communicate with the surface by minute pores which, in the living insect, are usually masked by an efflorescence of white waxy powder. In an article published in the *Entomologists' Monthly Magazine* (April, 1896, p. 85), I have drawn attention to the possible correlation between the circumgenital glands and the habit of oviposition. The following paragraph relating to the subject is quoted from that article:—

'It is a significant fact that, as far as I have at present observed (dealing with Ceylonese forms only), *those species of Diaspinæ that have no grouped glands are ovoviviparous, whilst those in which the wax-glands occur are strictly oviparous*. Of fourteen species of *Aspidiotus*, I find that nine are possessed of grouped wax-glands and lay eggs, while five species are glandless and produce living young. I have four species of *Aonidia*—all without the glands—and in all the embryo is fully developed before extrusion. The species of *Mytilaspis* occurring in Ceylon are amply provided with grouped glands, and deposit large numbers of eggs. In *Diaspis* the same conditions occur. In *Fiorinia* three oviparous species are provided with glands, and one ovoviviparous insect is without them. The same rule holds good in *Chionaspis*, with the doubtful exception of *Ch. biclavis*, Comst.'

Small groups of similar glands often occur round the openings of one or both pairs of stigmata. Here also the glands are productive of a powdery secretion, the purpose of which I imagine to be the formation of a protective covering to the spiracles, impervious to water while freely admitting the necessary air. Similar glandular pores occur on the stigmatic regions in the *Lecaniinæ*, accounting for the lines of white waxy matter that extend across the under surface of the body from the stigmatic clefts. The circumgenital glands may perhaps have a somewhat similar function, namely, the secretion of a waxy powder to protect the eggs.

The anal aperture is always found on the dorsal surface of the pygidium, but its position there varies greatly in different genera, and to a less degree in different species. As a general rule, we find that in *Aspidiotus* the aperture is nearer the extremity than in other genera, and always below the level of the genital orifice. At the opposite extreme comes *Mytilaspis*, in which the anus is usually near the base of the pygidium, considerably above the level of the genital orifice. In *Diaspis* the anal is below the genital aperture, but nearer to it than is usually the case with *Aspidiotus*. In *Chionaspis* and *Fiorinia* the anal is usually slightly above the level of the genital orifice. In *Aonidia*, in which the genital orifice is near the base of the pygidium, the anal aperture is placed about the middle of the segment. In doubtful cases, this character may perhaps assist in the determination of the genus.

The margin of the pygidium bears various processes that are found of great use in the differentiation of species. They are almost constant in the same species, while showing a great variation between distinct species. The most conspicuous of these processes are the prominent tooth-like chitinous lobes (*pl.* I. *fig.* 14, *i*) of which there may be from one to four pairs. The central pair sometimes coalesce to form a single median lobe. The lobes, together with the simple marginal prominences (from which it is sometimes difficult to separate them) seem to be employed in shifting obstacles, such as hairs and fibrous matter on the surface of the plant, that would otherwise interfere with the formation of the puparium. They are often strengthened by thickened ingrowths of the body-wall.

Between and beyond the lobes are the 'squames' ('plates' of Comstock, 'scaly hairs' of Maskell), which may be flattened and fimbriated as in *Parlatoria* and many species of *Aspidiotus* (*fig.* 16, *d, e, f*) or tapering and spiniform as in *Diaspis* and *Mytilaspis* (*fig.* 16, *a*), with numerous intermediate forms. The squames are concerned in the weaving of the scale and are associated with some of the tubular spinnerets described below.

There are also a few inconspicuous simple 'spines' (*fig.* 14, *m*) occurring usually in pairs, of which one is placed dorsally and the other ventrally. Each pair of these spines probably indicates a suppressed segment of the body.

The 'tubular spinnerets' are perhaps the most important organs connected with the pygidium, though they are far from being the most conspicuous. They are the principal if not the sole agents concerned in the secretion of the covering-scale. They are not confined to the pygidium, but occur in many species on the free abdominal segments also, either in groups on the lateral margin or extending in a transverse row across the segments. There are two principal forms noticeable: the 'cylindrical' which consists of a comparatively short and broad tube with parallel sides (*fig.* 14, *h*), and the 'filiform' which extends far into the body in the form of a narrow thread-like tube (*fig.* 14, *g*). Between these extremes may be distinguished a 'trumpet-shaped' tube (*fig.* 14, *f*), but these three types may be connected by many intermediate forms. They are all constructed on a similar plan. At the free inner extremity of the tube is a capitate organ, with a thickened rim and a central tongue or bulbous process. The tubes, having thin chitinous walls, remain visible in the skeletonised preparation, but the essential parts of the organ—the silk glands and their delicate ducts—are destroyed with the soft parts of the body. Dr. Antonio Berlese, in his most excellent paper referred to above, shows that the silk glands communicate by long delicate ducts with the capitate extremities of the tubular spinnerets. I take the liberty of copying one of his figures representing a spinneret with its glands (*fig.* 15). Some of the tubular spinnerets open on to the dorsal surface or the extreme margin by conspicuous oval or semilunar pores, others communicate with the squames, which are minutely perforate at their distal extremity.

The ovaries occupy the greater part of the body after gestation, extending even to the anterior margin, so that the whole insect sometimes appears to be tightly packed with the embryos.

For a description of the muscular and nervous systems I must again refer the student to Dr. Berlese's exhaustive work and admirable figures, in which the physiology of the insect is most minutely treated.

The first larval stage of the male insect is indistinguishable from that of the female, and the second does not greatly differ in the two sexes so far as the insects themselves are concerned, though the characters of the puparia diverge during this period. The male of the second stage is perhaps rather more oblong than the female at the same period. It is without limbs, the rostral apparatus is present, and the terminal segment is in the form of a pygidium, fringed with lobes and squames and provided with tubular spinnerets. Towards the end of this stage the future ocelli become apparent as dark diffused spots on each side of the head. The dorsal parts do not become indurated at the time of the second moult, but the thin skin is pushed off backwards, revealing the pupa (*pl.* II. *fig.* 1).

In this third stage the male insect has lost all likeness to the female. The mouth-parts have entirely disappeared, the limbs, antennæ, and wings are present in a rudimentary form, enclosed in their pupal sheaths, but lying free from the body. The divisions of the thorax and abdomen are more distinct. The extremity of the body is at first simply rounded, but subsequently a conical spike is developed which contains the genital sheath. The rudimentary limbs also increase in size during this period.

The adult male, after shedding the pupal skin, remains for

some little time beneath the protection of the puparium. In this cramped position it manages to fully expand its wings, as may be proved by raising the puparium shortly before the emergence of the insect. The winged male makes its exit backwards from the hinder part of the scale, and in so doing the delicate wings are drawn upwards over and in front of the head, but resume their natural position as soon as the insect has completely freed itself from its covering.

The adult male is very different to the female in its final stage. The main divisions of the body are better defined, though the boundaries of the head and the several parts of the thorax are still rather vague. The disposition of the various parts may be best explained by a reference to *pl.* II. (*figs.* 2, 7, 8), in which (*a*) represents the head, (*b*) the prothorax, (*c*) the mesothorax, (*d*) the metathorax, and (*e*) the abdomen. It will be noticed that the head and thorax have a backward tendency from above downwards (*fig.* 8). The head has the apex directed forwards; the hinder part widened and forming the so-called genæ or 'cheeks' (*fig.* 4, *c*), which may be easily mistaken for part of the prothorax. There are four large and conspicuous ocelli with prominent corneæ (*figs.* 4, *a* and 5, *a*); one pair on the upper surface immediately behind the antennæ, the second pair more centrally disposed on the under surface. The true eyes (*figs.* 4, *b* and 5, *b*) are minute, colourless, and inconspicuous, situated on the lateral margins of the head in the angle between the upper pair of ocelli and the genæ. In some species they seem to be entirely suppressed or reduced to a mere spot. The external mouth-parts are wanting, the insect taking no food during this stage.

The dorsal area of the prothorax (*fig.* 8, *b*) is small, but the ventral parts of this segment extend far backwards. The mesothorax (*fig.* 8, *c*), on the contrary, exhibits its greater area on the dorsal surface, where it is covered by a prominent shield-shaped scute (*fig.* 2, *f*) with a distinct transverse band (*fig.* 2, *g*) behind. This band is often of a darker colour than the rest of the insect, and is known as the 'apodema.' On the under surface of the mesothorax is a well-defined chitinous scute (*fig.* 7, *f*), broadest in front, with a median longitudinal division. This divided scute is always very distinct, being more highly chitinised and polished than the other parts of the ventral surface. The metathorax (*fig.* 8, *d*) bears on its upper surface the large shield-shaped 'scutellum' (*fig.* 2, *h*), the posterior margin of which is often partially overlapped by the softer parts of the segment. There is often a well-marked indentation on the lateral margin between the meso- and the metathorax. The division between the thorax and abdomen is unmarked by any constriction.

The abdomen consists of nine segments, including the conical piece giving rise to the genital spike. The full number can be distinguished on the dorsal surface; but on the under side of the abdomen one or more of the basal segments are concealed by the overlapping hinder margin of the metathorax.

A more or less distinct furrow runs along each side of the abdomen at a little distance from the margin on both the dorsal and ventral surfaces, separating off a well-defined marginal area.

The genital spike or sheath (*fig.* 10) is long and finely pointed, often half the length of the rest of the body. It consists of a single piece, with the edges incurved below to form an elongate valve, through which the exsertile penis can be extruded.

The antennæ (*fig.* 9) are normally of ten joints, as in all male Coccidæ. The first joint is short and broad, widest at the base. The second is usually more or less globular. The other joints are always rather elongate, and clothed with stout hairs. The terminal joint tapers abruptly to a blunt point at the apex, which bears a longish stout knobbed hair; and there are often two or more similar hairs springing from the side of this joint.

The single pair of wings spring from the sides of the mesothorax. They are long and ample (*fig.* 2), narrow at the base and broadly rounded at the tips, with a single nervure at the base dividing into two branches on the broader part of the wing. On the inner margin—close to the base—is a minute slightly thickened lobe, with a pocket which engages with the hooked extremity of the 'halter.' The membrane of the wing is hyaline and colourless in itself, though reflecting delicate iridescent tints. It is closely set with minute short hairs, which give it a slightly roughened appearance. When at rest, the wings lie in a horizontal position, completely overlapping each other above the back of the abdomen and extending considerably beyond the extremity of the body.

The 'balancers' or 'halters' (*fig.* 3) are attached to the lateral margins of the metathorax. There is a narrow flattened membranous portion (a) from the extremity of which, and at an acute angle, springs a stout hooked bristle (b), which engages with the pocket-shaped lobe at the base of the wing.

The legs (*fig.* 6) are long and moderately slender. The 'coxa' (*a*) is short and conical. The 'trochanter' (*b*) is rather long and roundly dilated at its distal extremity, which usually bears a single longish hair. The 'femur' (*c*) is broadly overlapped by the extremity of the trochanter. The 'tibia' (*d*) is widest at its distal extremity and set with some stout hairs. The single-jointed 'tarsus' (*e*), wide at the base, tapers rather abruptly to a point at its extremity, and is also set with stout hairs, particularly on the inner margin. The single 'claw' (*f*) is longish and fully pointed. There are from two to four conspicuous knobbed hairs or 'digitules' on the foot. When the full number are present, two of them (*h*) spring from the outer margin of the tarsus, and two (*g*) from the base of the claw on the inner side; but in many species one, or sometimes two, of the digitules are missing. The number of digitules in the foot of the male will be found a useful specific character. The two sets of digitules possibly represent suppressed tarsal joints. There is usually a considerable interval between the first and second pair of legs, owing to the elongate prosternum.

In some few species of *Diaspinæ* an apterous form of the male occurs. In these the parts of the body may be modified, the mesothorax not requiring the muscular development and highly chitinised scutes present in the winged forms. The antennæ also in the apterous forms may show modification, the joints being often shorter and less distinct, or one or more joints may be suppressed.

In the latest catalogue of Coccidæ (Prof. Cockerell's 'Check List'), the *Diaspinæ* are divided into fifteen classes—*Aspidiotus, Comstockiella, Diaspis, Aulacaspis, Pseudoparlatoria, Parlatoria, Syngenaspis, Mytilaspis, Pinnaspis, Chionaspis, Leucaspis, Ischnaspis, Fiorinia, Poliaspis,* and *Aonidia*. Twelve of these have been incorporated into the following synopsis. I am not sufficiently acquainted with the characters of the other three, *Aulacaspis, Pseudoparlatoria,* and *Pinnaspis,* to enable me to separate them satisfactorily; they appear to be offshoots of the genera *Diaspis, Aspidiotus,* and *Mytilaspis* respectively.

Those genera not yet recorded from Ceylon are included in round brackets ( ).

It is difficult with genera that approach each other so closely as do those of the *Diaspinæ* to give concise rules by which they can be confidently separated. The generic characters put forward in the following arrangement must be considered elastic—within reasonable limits. Abnormal forms will be found in nearly every genus.

SYNOPSIS OF GENERA.

A. Male puparium similar in general form and structure to that of the female.
- α. Pellicles of female completely superposed, and in both sexes more or less centrally situated.
  - (i) Female puparium sub-circular. Pellicles surrounded by a broad secretionary supplement. Circumgenital glands in not more than five groups. Male puparium similar to female, but more oblong. (*Pl.* I. *fig.* 7.)
    ASPIDIOTUS.
  - (ii) Puparia as in preceding genus. Circumgenital glands in more than five groups .... (COMSTOCKIELLA.)
  - (iii) Female puparium occupied almost completely by the greatly enlarged second pellicle with little or no secretionary supplement. Adult female reduced in size and enclosed within the second pellicle. Male puparium sub-circular; the pellicle surrounded by a broad secretionary supplement. (*Pl.* I. *fig.* 2)...... AONIDIA.
- β. Pellicles of female overlapping, and in both sexes placed at or close to the anterior extremity.
  - (iv) Female puparium broadly elliptical. Second pellicle large, with a moderate secretionary supplement. Pygidium with a continuous marginal series of broad fimbriated squames and large semilunar pores. Circumgenital glands in four groups. Male puparium irregularly elliptical; rather depressed. (*Pl.* I. *fig.* 6.)
    (PARLATORIA.)
  - (v) Female puparium elongate; with a moderately large secretionary supplement behind. Margin of pygidium as in preceding genus. Circumgenital glands in five groups. Male puparium smaller and narrower.
    (SYNGENASPIS.)
  - (vi) Female puparium usually elongate and narrow. Circumgenital glands in five groups. Male puparium similar to that of female, but smaller; the hinder extremity often with a hinge-like structure to allow of the easy exit of the adult insect. (*Pl.* I. *fig.* 5.) MYTILASPIS.

(vii) Female puparium long and very narrow; with parallel sides. Male puparium similar but smaller. Pygidium of female with a conspicuous coarse reticulation on the dorsal surface .... (ISCHNASPIS.)

(viii) Female puparium rather elongate; dilated behind. Male puparium smaller; with parallel sides. Circumgenital glands in two concentric series .... (POLIASPIS.)

(ix) Female puparium elongate: occupied almost entirely by the large second pellicle which encloses the adult female and is itself usually concealed by a covering of opaque white secretion. Male puparium similar in form but smaller. Circumgenital glands confluent, forming an irregular arch. Margin of pygidium with a continuous fringe of spine-like squames.
(LEUCASPIS.)

B. Male puparium white, elongate, narrow, with subparallel sides; usually with more or less prominent longitudinal carinæ: form and structure irrespective of or dissimilar to that of the female.

a. Pellicles of female overlapping.

(x) Female puparium sub-circular. Pellicles situate within the margin, a little to one side. (*Pl.* I. *fig.* 4.) Circumgenital glands usually in five groups ... DIASPIS.

(xi) Female puparium consisting principally of the enlarged second pellicle which completely encloses the adult insect and the eggs. (*Pl.* I. *fig.* 3.) Circumgenital glands usually in five groups of which the upper three are often confluent .... FIORINIA.

(xii) Female puparium elongate or elliptical: usually dilated behind. (*Pl.* I. *fig.* 1.) Circumgenital glands usually in five groups .... CHIONASPIS.

ASPIDIOTUS, Bouché.

Species in which the puparium of the female is normally more or less circular; occasionally oval (*cyanophylli*), or elongated (*inusitatus*). Pellicles superposed, approximately central. In this genus the first larval pellicle consists of the dorsal parts of the skin only, the ventral parts, with the antennæ and limbs, being completely separated and incorporated into the ventral scale of the puparium. This character distinguishes *Aspidiotus* from all other genera with which I am acquainted, the first pellicle in these other genera having the antennæ attached to the anterior margin, while the remains of the limbs will be found beneath the posterior extremity. Ventral scale usually very delicate, a mere film, adhering to the plant; but in some species (*camelliæ, aurantii, &c.*) it is of stouter texture and remains attached to the upper parts; while in others the marginal area only may be thickened.

Male puparium similar in structure to that of the female; but usually smaller and more oblong.

Adult female oval; broadly rounded in front, narrowing behind. Divisions of abdominal segments often indistinct. After oviposition the abdominal parts usually become greatly contracted, completely altering the form of the insect. Anal aperture situated nearer the extremity than the genital aperture. Circumgenital glands usually in four groups; a few species with five groups; others with none. Tubular spinnerets cylindrical, filiform, or trumpet-shaped. An aonidiform stage is observable in some species at the time of the second moult (*vide pl.* XV. *fig.* 4).

Adult male rather broad; moderately depressed. A prominent colourless tubercle on lateral margin of head representing the rudimentary eyes. Antennæ usually with three knobbed hairs on the terminal joint.

For convenience of synopsis the Ceylon species of *Aspidiotus* may be divided into primary groups according to the number of lobes on the pygidium; these groups may be subdivided according to the circumgenital glands.

Synopsis of Species.

A. Pygidium with 8 lobes.
    (a.) Circumgenital glands in four groups.
        (i.) Puparium rather large, flat, pale brown. Pellicles yellow. Female insect clear shining brown with deep transverse groove across thorax .... *trilobitiformis*.

B. Pygidium with 6 lobes.
    (a.) Circumgenital glands in four groups.
        (ii.) Puparium olive brown; pellicles clear, fulvous. Female insect whitish or yellow .... *ficus*.

  (iii.) Puparium brown; pellicles blackish. Female insect purplish .... *rossi*.
  (iv.) Puparium opaque, fulvous or brownish, resembling the bark of the plant. Female insect yellow .... *osbeckiæ*.
  (v.) Puparium transparent, colourless or greyish; pellicles yellow. Female insect yellow .... *lataniæ*.
  (vi.) Puparium semi-transparent, oblong oval. Female insect pale yellow .... *cyanophylli*.
  (vii.) Puparium semi-transparent, irregularly lobed. Female insect yellow .... *excisus*.
 (b.) Circumgenital glands wanting.
  (viii.) Puparium fulvous, flat, covering small pits in the leaf in which the insect lies. Female insect yellow .... *putearius*.
  (ix.) Insect occupying minute galls on the leaf. Female insect yellow .... *occultus*.
  (x.) Puparium semi-transparent, reddish fulvous, adhering to the insect. Female insect orange yellow .... *aurantii*.
C. Pygidium with two lobes.
 (a.) Circumgenital glands in four groups.
  (xi.) Puparium whitish, or coloured by the admixture of fibrous particles of the plant. Female insect yellow .... *cydoniæ*.
 (b.) No circumgenital glands.
  (xii.) Puparium whitish, fulvous, usually oblong, with pellicles towards one extremity. Female insect yellow .... *camelliæ*.
D. Pygidium with one (median) lobe.
 (a.) Circumgenital glands in two groups.
  (xiii.) Insect imbedded between the tissues of the plant. Female insect yellow .... *secretus*.
E. Pygidium without lobes.
 (a.) No circumgenital glands.
  (xiv.) Puparium elongated; fulvous; dorsal and ventral sides similar. Female insect reddish yellow .... *inusitatus*.

ASPIDIOTUS TRILOBITIFORMIS, *Green.*
(PLATE IV.)

Aspidiotus trilobitiformis, Green, 'Catalogue of Coccidæ,' *Ind. Mus. Notes,*
Vol. IV. No. 1 (1896).

Female puparium large, almost flat; usually semicircular (*fig.* 3) or deltoid from arrest of growth by prominent veins of the leaf, seldom circular, the insect apparently preferring a position close to the midrib of the leaf (*fig.* 1). Colour, pale reddish brown. Pellicles yellow, the second usually depressed. Diameter of puparium, 3 to 4·50 mm.

Male puparium unknown.

Adult female (*figs.* 4, 5) clear brown; surface hard and horny, polished, with numerous delicate transverse striated lines. Form oblong, rounded in front, tapering to a point behind; dorsal surface flattened; ventral surface slightly tumid; segments distinct and strongly defined; a deep transverse groove on dorsal surface between the prothoracic and mesothoracic segments; a large irregular depressed space on each side of rostrum, covered with white waxy secretion, marking the position of the parastigmatic glands, of which there is a group consisting of from 12 to 20 orifices in front of each of the anterior stigmata. Pygidium (*fig.* 6) with eight prominent obscurely tricuspid lobes; mesal pair stoutest, but scarcely as long as second; others rather slender. Squames deeply fringed; two in the mesal and first spaces, and three in the second and third spaces between the lobes. Lateral margin of pygidium irregularly serratulate with two deep notches marking the position of the obliterated second and third abdominal segments. On the dorsal surface is an extensive reticulated tract completely occupying the median area of the pygidium between the base and the anal aperture, the boundaries well defined and constant, the spaces of irregular size and shape, crowded together, and forming a pattern not unlike that of crocodile leather. Circumgenital glands in four groups; orifices numerous, upper laterals with 21 to 24, lower laterals with 16 to 27; in every case the upper laterals contain the larger number of orifices; in one specimen were two single separate orifices in the place of an anterior median group. Tubular spinnerets of the filiform type, opening on the dorsal surface by large conspicuous pores arranged in definite linear series; the ducts themselves very delicate and difficult to trace. Similar pores and spinnerets on the other abdominal segments. Genital aperture between the lower lateral gland groups. Anal aperture about half way between extremity and genital opening. Length 1·50–1·80 mm. Breadth about 4 mm.

Adult male unknown.

Egg (*fig.* 2) pale purplish.

Habitat on under surface of leaves of *Dalbergia Championii*, and of an unidentified tree. Punduloya; (August).

Very closely allied to (possibly only a variety of) *A. theæ*, Maskell, found on stems of tea plants in Assam. The latter has a more convex puparium, the body of the female is comparatively broader, and the lateral lobes of the pygidium are proportionately much smaller than in *A. trilobitiformis*. Allied also to *A. eucalypti*, Maskell; *A. articulatus*, Morgan; and *A. rufescens*, Cockerell, which form a section characterised by the distinct separation of the prothoracic from the mesothoracic segments.

EXPLANATION OF PLATE IV.

ASPIDIOTUS TRILOBITIFORMIS.

(All figures, except No. 1, more or less enlarged.)

Fig. 1. Insects, nat. size, on leaf.
    2. Egg.
    3. Female puparium.
    4. Adult female, dorsal view.
    5.   ,,   ,,   ventral view.
    6. Pygidium of adult female, dorsal view.

ASPIDIOTUS FICUS (*Riley*), *Comstock.*
(PLATE V.)

Chrysomphalus ficus, Riley, MSS., Ashmead, *American Entomologist*, 1880, p. 267.

Aspidiotus ficus, Comstock, *Canadian Entomologist*, Vol. XIII. p. 8.

Female puparium circular, moderately convex, smooth; dark olivaceous brown or reddish brown, paler at margin (*fig.* 2). Pellicles reddish yellow, always partially obscured by a layer of secretion which is reddish brown above the first, and pale olivaceous above the second pellicle. In the centre a circular raised disc is usually exposed, the secretionary covering being here worn off. In young specimens the centre is covered by a raised patch of opaque white secretion. The first pellicle convex above; the second often slightly concave; the form may best be observed from the inside of the scale (*fig.* 3) where the exuviæ are more fully exposed. Ventral scale obsolete. Diameter 1 to 2 mm.

Male puparium, unobserved in Ceylon; but, on a plant of *Garcinia cambogia*, labelled from Ceylon, growing in one of the plant houses at Kew Gardens, I found a large colony of these insects containing both male and female scales. The male puparium (*fig.* 6) is dark brown with pale grey margin. Pellicle reddish fulvous. Length, 0·80 mm.

Adult female yellow, or white mottled with yellow. Body broadly rounded in front, tapering suddenly to a point behind (*fig.* 7). On the margin of the mesothorax is a small thickened patch bearing a stout thorn-like spine (*figs.* 7 and 8). Pygidium (*fig.* 4) with six prominent lobes sub-equal in size, each notched on the lateral edge. At a short distance beyond the lobes the lateral margin is thickened and projecting, with minute serrations and two deep indentations. Squames deeply fringed; two in the mesal and first spaces, three in the second space, and three beyond the third lobe, these last being bifurcate and fringed on their lateral edges. Circumgenital glands in four groups; lower laterals with two to four, upper laterals with six to eight orifices. A large number of conspicuous tubular spinnerets, varying from the filiform to the trumpet-shaped type, some opening by inconspicuous dorsal pores in two short series on each side, others opening on to the margin near the extremity. Anal aperture small, close to extremity. Genital aperture between the upper and lower groups of glands. Length 0·80 to 1 mm.

Adult male unobserved in Ceylon. Said by Comstock to be orange-yellow in colour with dark brown apodema.

Eggs and young larvæ, yellow.

Habitat in Ceylon, on upper surface of leaves of *Rhododendron arboreum* (*fig.* 1). Newera Eliya (January).

This species has been very carefully studied by Prof. Comstock in America, where it attacks orange trees. It is said to be very injurious to the younger trees, and to greatly disfigure the fruit. Prof. Comstock found that the insect required about sixty days to develop from the egg to the adult. He considers the species one to be feared, and recommends orange-growers to eradicate the pest before it becomes unmanageable.

EXPLANATION OF PLATE V.

ASPIDIOTUS FICUS.

(All figures, except No. 1, more or less enlarged; figs. 6, 7, and 8 are drawn from specimens collected at Kew on Garcinia cambogia.)

Fig. 1. Piece of rhododendron leaf, with insects *in situ*, nat. size.
    2. Female puparium, from above.
    3.   ,,   ,,   ,,   below.
    4. Pygidium of adult female.
    5. Female puparium, more advanced specimen, from below.
    6. Male puparium, from above.
    7. Adult female, ventral view, showing spine on side of thorax.
    8. Thorn-like spine, from margin of mesothorax.

ASPIDIOTUS ROSSI, Maskell.
(PLATE VI.)

Aspidiotus Rossi, Maskell, *Trans. N. Z. Instit.*, 1891, p. 11.

Female puparium circular (*fig.* 2), or irregularly oblong (*fig.* 4), flattish, opaque, reddish brown or dark brown; inner surface darker, almost black. Pellicles blackish, frequently obscured by a layer of brownish secretion, with central boss and concentric ring; sometimes depressed, sometimes slightly elevated. Ventral scale obsolete, a white powdery film on surface of leaf, except at margins, where it is stouter and adheres to the dorsal scale (*fig.* 4). Diameter, 2 to 3 mm.

Male puparium stated by Maskell to be smaller and lighter in colour than that of the female. I have not found the male insect in Ceylon.

Adult female (*fig.* 3) broadly pyriform, terminal segment tapering suddenly to a point; median area tumescent; margins flattened. Colour of living insect at first milky white or ochreous, tinged with purplish, which deepens with age and extends over the greater part of the thorax, the flattened marginal area and the abdominal segments remaining ochreous. Colour of dead and dried insect, brownish yellow. Stigmata conspicuous; no parastigmatic glands. Pygidium (*fig.* 5) with six prominent, obscurely tricuspid lobes, all well developed and sub-equal in size; margin beyond the lobes, with seven projecting tooth-like processes, forming a bold and regular serration; margin between the lobes, squarely but not deeply incised. Squames deeply fringed, two between median lobes, two between first and second, three between second and third, and one or two in the space between the third lobe and the first marginal prominence. Circumgenital glands in four groups; upper laterals with 9 to 12 orifices, lower laterals with 8 to 9, their position indicated in the living insect by the presence of four white waxy patches. A large number of very delicate filiform tubular spinnerets, opening on the dorsal surface by small and rather inconspicuous pores arranged in definite linear series running upwards from the margin. Larger cylindrical or trumpet-shaped ducts nearer the extremity, opening on the margin between the lobes. Anal aperture slightly caudad of the lower spinneret groups. Length about 1·50 mm.

Adult male unknown.

Eggs pale purplish. Hatched shortly after extrusion. Well-developed embryos can be seen within the body of the parent insect.

Young larvæ very pale reddish, broadly oval; caudal setæ short.

Habitat in Ceylon on under-surface of leaves of *Capparis Moonii* (*fig.* 1). Originally reported from Australia, where it is said to occur on various plants and trees.

EXPLANATION OF PLATE VI.

ASPIDIOTUS ROSSI.

(*All figures, except No. 1, more or less enlarged.*)

Fig. 1. Leaf of *Capparis Moonii*, with insects *in situ*, nat. size.
    2. Female puparium, from above.
    3. Adult female, ventral view.
    4. Female puparium, from below.
    5. Pygidium of adult female.

ASPIDIOTUS OSBECKIÆ, Green.
(PLATE VII.)

Aspidiotus osbeckiæ, Green, 'Catalogue of Coccidæ,' *Ind. Mus. Notes*, Vol. IV No. 1 (1896).

Female puparium convex; normally circular (*fig.* 9), but by crowding the scales often become oblong (*fig.* 10), or of irregular outline. Colour brownish, paler at margin, opaque; tint varying with that of the bark upon which it rests, and partly due to the incorporation of loose fibrous matter from the bark itself. Pellicles slightly eccentric; yellow or reddish, usually obscured by a layer of secretion; a small boss in centre of first pellicle. Ventral scale obsolete, appearing as a whitish scar on the bark after the removal of the insect. Diameter about 1·50 mm.

Male puparium (*fig.* 6) similar to that of female, but more oblong; breadth less than half length. Size averaging 1·30 by 0·50 mm.

Adult female (*figs.* 11 to 14), pale yellow, darkening with age; pygidium, tinged with reddish brown. Before gestation broadly oval; margin more or less distinctly lobed; abdominal segments well defined. After gestation the abdominal segments contract, carrying the pygidium upwards, until it is overlapped by the sides of the thorax (*fig.* 13). Antennæ rudimentary, but more conspicuous than in many species, the tubercle being very prominent, unequally dilated at extremity, with a depression near the summit and a stout curved hair from the base (*fig.* 16). No parastigmatic glands. Pygidium (*fig.* 15) with six tricuspid lobes; mesal lobes largest; third pair very small; margin between lobes broadly incised. Squames deeply incised and fringed, two between each lobe, and two to four beyond, the latter sometimes almost obsolete. Spines at base of each lobe, the dorsal ones largest. A complete marginal series of small hairs all round the body. Circumgenital glands in four or five groups, the anterior group being absent or represented by one (rarely two) orifices, anterior laterals with six to nine, and posterior laterals with three to five orifices. On each side of pygidium are two series of delicate filiform tubular spinnerets opening by distinct oval pores on the dorsal surface, a few opening by the margin between the lobes. Similar filiform spinnerets on the lateral margins of all the abdominal segments. Length 1 to 1·50 mm.

Adult male (*fig.* 12) orange yellow; apodema dark brown or black. Ocelli black; lower pair largest, central, narrowly separate (*fig.* 2), a small colourless prominent tubercle on each side representing the true eye. Antennæ hairy; joints increasing in length to sixth, which is longest, then decreasing to tenth; terminal joint tapering, with a stout knobbed hair at apex, and two fine knobbed hairs at side (*fig.* 3). Foot with four digitules (*fig.* 4); tarsus shorter than tibia. Genital spike nearly half length of body. Total length of insect, nearly 1 mm.

Eggs yellow; hatched almost immediately after extrusion.

Young larva with two small lobes at extremity of body (*fig.* 7); caudal setæ short; colour bright yellow.

Habitat on stems of *Osbeckia*, Pundaloya (August). Frequently in company with *Asp. camelliæ* and *Chionaspis biclavis*.

A most inconspicuous species and difficult to detect, the colour of the scales almost exactly resembling that of the bark upon which they rest.

EXPLANATION OF PLATE VII.

Aspidiotus osbeckiæ.

(All figures, except No. 8, more or less enlarged.)

Fig. 1. Adult male, from above.
    2. ,, ,, head, from below.
    3. ,, ,, terminal joint of antenna.
    4. ,, ,, foot.
    5. ,, ,, side view.
    6. Male puparium.
    7. Young larva.
    8. Branch of *Osbeckia* with insects, nat. size.
    9. Female puparium, from above.
  10. ,, ,, from below (distorted example).
  11. Adult female, before oviposition, dorsal view.
  12. ,, ,, ,, ,, ventral view.
  13. ,, ,, after oviposition, dorsal view.
  14. ,, ,, ,, ,, ventral view.
  15. Pygidium of adult female, from below.
  16. Antenna of adult female.

ASPIDIOTUS LATANIÆ, *Signoret.*
(PLATE VIII.)

Aspidiotus lataniæ, Sign. *Essai*, 1869, p. 124.
Aspidiotus transparens, Green, *Insect Pests of the Tea Plant*, 1890, p. 22.

Female puparium circular, flattish; no concentric lines of growth, but fresh specimens with fine radiating wrinkles; secreted area transparent, whitish or colourless, showing distinctly the sublying insect and eggs (*figs.* 3, 4). Pellicles very pale yellow, transparent, nearly central. Ventral scale represented only by a delicate film on surface of leaf. Diameter 1·50 mm.

Male puparium (*fig.* 2) similar to that of female, but smaller, oval, and with single pellicle. Size 1 mm. by 0·75 mm.

Adult female (*figs.* 6, 7) bright pale yellow; pygidium tinged with brown towards the extremity, four white waxy patches indicating externally the circumgenital glands. Body pyriform, tapering and pointed behind; thoracic area broadly rounded; anterior margin straight or slightly concave, with often a minute prominence on each side of the depression (*fig.* 6). Antennæ rudimentary, a minute tubercle with curved hair at base. No parastigmatic glands. Pygidium (*fig.* 10) with six prominent lobes; median pair stoutest, obscurely tricuspid, dark-coloured, scarcely as long as second pair; second and third pairs slender, very delicate and transparent, contracted at base and notched on outer side; each lobe situated on a prominent point of the margin. Margin coarsely serrate for some distance laterad of the lobes. Two simple or slightly divided squames between the median lobes; two, deeply fringed, between median and second lobes; three between second and third lobes; and a series of six or seven, fringed on their outer sides, laterad of the third lobe. A pair of small spines at base of each lobe; a fourth pair on each side on a prominent point shortly beyond the last lobe; a fifth in a small cleft immediately beyond the last squame; and a sixth at base of pygidium. Circumgenital glands in four groups; upper laterals with from six to eleven; lower laterals with from four to six orifices. Tubular spinnerets of the filiform type, long and delicate, but unusually conspicuous, opening on to the extreme margin (*fig.* 11); nearer the extremity are broader trumpet-shaped ducts, also opening on to the margin. Length 0·75 to 1 mm. Breadth 0·50 to 0·75 mm.

Adult male reddish (*fig.* 5), or pale yellow with reddish apodema (*fig.* 5, *a*); the darker variety was bred from specimens occurring on *Dalbergia*, the paler on tea. Body depressed, rather broad, abdomen short. Genital spike long and slender, nearly half length of body. Ocelli black; upper pair marginal; lower pair central and contiguous (*fig.* 14); a minute colourless tubercle on each side immediately behind the upper ocelli represents the true eye. Antennæ hairy; nearly as long as body; two basal joints short; third joint longest; subsequent joints gradually decreasing in length to ninth; tenth a little longer than ninth, tapering in front, a stout sh knobbed hair at its apex and two finer ones on the side (*fig.* 13). Foot (*fig.* 12) with two knobbed hairs (one on claw and one on tarsus). Tarsus two-thirds length of tibia ; tibia nearly equal to femur. Wings ample, hyaline. Total length 0·75 mm.

Eggs pale yellow (*fig.* 8), disposed in a ring round the body of the parent (*fig.* 4).

Young larva (*fig.* 9) pale yellow. Eyes minute, black. Feet with four knobbed hairs. Posterior extremity fringed with short hairs and papillæ. Caudal setæ short, about one quarter length of body.

Habitat in Ceylon on various species of palm, on tea, *Loranthus*, *Dalbergia championii*, and several unidentified shrubs ; occupying usually the under surface only of the leaves ; both sexes crowded together in large clusters (*fig.* 1). Very widely distributed in all parts of Ceylon. I have seen it in distinctly injurious numbers on the fronds of various ornamental palms. Its effect upon the tea plant is fortunately not so marked, though it sometimes occurs in vast numbers on individual plants, usually affecting the older leaves only. Type described from specimens found on *Latania*, an African palm.

I refer this insect rather doubtfully to this species. It differs in several points from the type of *Aspidiotus lataniæ* as described by Signoret, principally in the number of orifices in the spinneret groups. The conspicuous hair-like tubular spinnerets are made by Signoret a specific character. I at first thought this insect was the well-known 'Oleander scale'—*Aspidiotis nerii ;* but careful comparison with typical specimens shows me that it cannot be that species. The character of the tubular spinnerets is very different in the two species, *A. nerii* having short cylindrical ducts only.

EXPLANATION OF PLATE VIII.

ASPIDIOTUS LATANIÆ.

(All figures, except No. 1, *more or less enlarged.)*

Fig. 1. Leaf, with insects *in situ*, natural size.
    2. Male puparium.
    3. Female puparium.
    4.   ,,   ,,   showing disposition of eggs.
    5. Adult male.
    5A.   ,,   pale variety from tea plant.
    6. Adult female, dorsal view.
    7.   ,,   ventral view.
    8. Egg.
    9. Young larva.
   10. Pygidium of adult female.
   11. Diagram showing tubular spinnerets opening on margin of pygidium.
   12. Foot of adult male.
   13. Terminal antennal joint of adult male.
   14. Diagram of head of male, from below, showing eyes and ocelli.

ASPIDIOTUS CYANOPHYLLI, *Signoret.*
(PLATE IX.)

Aspidiotus cyanophylli, Sign., *Essai*, 1869, p. 119.

Female puparium at first rounded oval, afterwards very oblong oval (*fig.* 2), one side often compressed by contact with a prominent vein of the leaf. A convex median area separated rather distinctly from a flatter marginal zone; transparent, colourless, or very slightly stained with ochreous, marked with fine concentric lines of growth. Pellicles approximately central, transparent bright yellow. In middle of first pellicle is a circular elevated disc with a white central boss and radiating raised lines (*fig.* 3). Ventral scale very thin and delicate, a mere film on surface of leaf. Length 1·50 to 2 mm. Breadth 0·75 to 1 mm.

Male puparium said by Signoret to be similar to that of female, but more oblong.

Adult female (*fig.* 4) oblong, unusually so for insects of this genus; margin of cephalic extremity flattened; segments moderately distinct. Antennae (*fig.* 5) rudimentary. No parastigmatic glands. Pygidium (*fig.* 6) with six tricuspid lobes; median pair much larger and darker than the others, roundly tricuspid; the second and third pairs slender and pointed; the third sometimes almost obsolete. Three deep marginal notches on each side (one laterad of each lobe), more apparent on dorsal surface, where they extend further inwards. Squames long and deeply fringed, two between median lobes, two between first and second lobes, three between second and third, and four or five laterad of the third lobe. Circumgenital glands in four groups; orifices few, upper laterals with 3 to 5, lower laterals with 5 or 6. Tubular spinnerets filiform, opening by long ducts on to the extreme margin; a few (about three on each side) opening by inconspicuous dorsal pores. Anal aperture near extremity. Genital aperture near the centre of the pygidium. Length from 1 to 1·50 mm. Breadth 0·50 to 0·75 mm.

Adult male unknown. Signoret describes the male puparium, but makes no mention of the adult.

Eggs greenish yellow.

Habitat in Ceylon on tea and cinchona plants, also on a species of palm (*fig.* 1); usually confined to the under surface of the leaves. Larger colonies occur on the fronds. A still more elongated form (*fig.* 7) occurs on *Cycas* at Kandy, crowded upon upper and under surface of fronds. Signoret's type was described from specimens found on a Venezuelan plant (*Cyanophyllum magnificum*) in the Luxembourg Gardens. Comstock has found it in the United States on several species of *Ficus*.

EXPLANATION OF PLATE IX.

ASPIDIOTUS CYANOPHYLLI.

(All figures, except No. 1, more or less enlarged.)

Fig. 1. Piece of palm leaf, with insects, nat. size.
 2. Female puparium.
 3. First larval pellicle, showing disc and boss.
 4. Adult female, ventral view.
 5. ,, ,, rudimentary antenna.
 6. ,, ,, pygidium, from below.
 7. ,, ,, elongated form from Cycas.
 8. Female puparium, under side, from Cycas

ASPIDIOTUS EXCISUS, sp. nov.
(PLATE X. figs. 7-9.)

Female puparium (fig. 8) convex, of irregular outline, margin often lobed, thin, and semi-transparent, whitish or very pale ochreous; bearing many long hairs, which have been separated from the plant and incorporated into the scale. Pellicles yellow, approximately central. Ventral scale obsolete. Diameter about 1 mm.

Male puparium (fig. 7) smaller and more oblong, similar in structure to that of female. Size, 0·80 by 0·50 mm.

Adult female yellow; broadly pyriform. No parastigmatic glands. Pygidium (fig. 9) appearing truncate from the fact that the extremity is broadly and deeply excised. Lobes six; median pair largest, but sunk in a deep, squarely cut recess, which extends on each side as far as the second lobe, their distal extremities plain or very slightly emarginate and slanting; second and third lobes small but prominent, the second projecting distinctly beyond the mesal lobes. Squames deeply fringed, two in the mesal and first spaces, three in the second space, and four beyond the third lobe. Small spines at base of each lobe, and three at intervals along the lateral margin. Circumgenital glands in four groups; upper laterals with 8 to 15, lower laterals with 7 to 9 orifices. A few long filiform tubular spinnerets, opening on to the lateral margin; others nearer the extremity shorter and trumpet-shaped; four or five short cylindrical ducts on each side of anal aperture, opening by large but inconspicuous pores on the dorsal surface, their free ends directed downwards. Genital aperture between lower gland groups. Anal aperture nearer the extremity. Length 0·60 to 0·80 mm.

Adult male unknown.

Eggs numerous, yellow.

Habitat on under surface of leaves of *Cyanotis pilosa*. Punduloya (December).

A small but inconspicuous species, but easily distinguished by the square recess at extremity of pygidium.

EXPLANATION OF PLATE X.
ASPIDIOTUS EXCISUS.
(All figures more or less enlarged.)

Fig. 7. Male puparium.
    8. Female puparium.
    9. Pygidium of adult female.

ASPIDIOTUS PUTEARIUS, sp. nov.

(PLATE X. figs. 1-6.)

Female puparium (fig. 2) round, flat, or slightly concave, forming an operculum to the pit-like depression in which the insect rests. Colour very pale brownish ochreous, semi-opaque; minutely rugose, with concentric lines of growth. Pellicles central, pale yellow; the second pellicle slightly concave, the first slightly convex. Ventral scale obsolete, a mere powdery film lining the cavity below the insect. Diameter 1·50 mm.

Male puparium (fig. 6) broadly oval; similar in texture to that of the female; median area convex where it covers the insect. Size, 1·12 × 1 mm.

Adult female (fig. 3) bright pale yellow, terminal segment colourless. Almost circular in outline; flat or slightly concave above; highly convex below, where it fits into cavity in the leaf. Segments well defined above, inconspicuous below. Eyes minute, black, sub-marginal, seen only from above. Antenna rudimentary, a small tubercle with curved hair at base. A few small hairs at regular distance round margin of body. No parastigmatic glands. Pygidium comparatively small. Lobes, six; median pair largest, obscurely tricuspid; others smaller, very delicate, and transparent, third usually emarginate on outer side. Deeply fringed squames between the lobes. Margin laterad of lobes with four or five fringed squamous marginal processes. Margin dorsad of second and third lobes thickened, and produced into prominent points. Small spines at base of each lobe; a fourth and fifth pair further up on each side. Anal aperture about half way between genital aperture and extremity. No circumgenital glands. A large number of tubular spinnerets in four or five diagonal series on each side converging towards extremity of pygidium; each spinneret consisting of a short cylindrical tube, with thickened rim and tongue-like process at free end, opening on the dorsal surface by well-defined oval pores; those nearest the extremity opening apparently on to the margin. Diameter about 1 mm.

Adult male reddish. Foot (fig. 5) with four-knobbed hairs, tarsus shorter than tibia, Antennæ normal. (Described from dead, shrivelled specimens.)

Habitat on *Strobilanthes viscosus*, forming and occupying small pits on under surface of leaves; corresponding discoloured prominences, like half-formed galls, appearing on the upper surface (fig. 1). Not common, but occurring on individual plants in enormous numbers, covering every leaf. Pundaloya (January and May).

Mr. Maskell has described (*Trans. N. Z. Instit.*, 1890, p. 10), under the name of *Aspidiotus fodiens*, an Australian insect of similar habits, forming pits in the leaves of a species of acacia. That insect, however, differs from *A. putearius* in several important points, notably in the possession of circumgenital glands.

EXPLANATION OF PLATE X.

ASPIDIOTUS PUTEARIUS.

(All figures, except No. 1, more or less enlarged)

Fig. 1. Leaf of Strobilanthes, with insects, nat. size.
    2. Female puparium.
    3. Adult female, dorsal view.
    4. Pygidium of adult female, ventral view.
    5. Foot of adult male.
    6. Male puparium.

ASPIDIOTUS OCCULTUS, Green.

(PLATE XI.)

Aspidiotus occultus, Green, 'Catalogue of Coccidæ,' *Ind. Mus. Notes*,
Vol. IV. No. 1 (1896).

A very small species, forming minute rounded galls on the upper surface of leaves of *Grewia orientalis* (*fig.* 5). Although the galls appear on the upper surface, the insects are in reality placed on the under surface, at first forming depressions as in *A. putearius*, but in the present species the process is carried further by the formation of a cell, which almost completely encloses the insect. Occasionally two individuals occupy a single cell.

Female puparium represented externally by an ochreous scale closing the aperture of the gall (*fig.* 6). A whitish film lining the cavity may be taken to represent the ventral scale. The pellicles do not appear on the surface, but lie within the gall (*fig.* 7). The second pellicle is very large, sometimes as large as, or even larger, than the adult insect, in this respect approaching the genus *Aonidia*. The aperture of the gall, opening on the under surface of the leaf, is surrounded by a prominent irregularly lobed rim (*figs.* 6 and 7). Diameter of gall from 0·50 to 0·75 mm.

Male puparium occupying shallow depressions on the under surface of the leaf; oval, flat; transparent, very pale reddish yellow; pellicle, bright yellow (*fig.* 8). Size about 1 × 0·75 mm.

Adult female broadly pyriform; flattish or slightly concave above, median dorsal area distinctly depressed (*fig.* 10); highly convex below, almost hemispherical (*fig.* 9). The insect usually lies sideways in the gall (*fig.* 7). Colour pale yellow; terminal segment brownish, with the median lobes dark brown. Eyes minute, black, on dorsal surface. No parastigmatic glands. Anterior stigmata situated very far forward. Pygidium with six lobes; the median pair long and prominent, tricuspid; others much smaller, emarginate on their outer sides. Long and deeply fringed squames between and beyond the lobes. No circumgenital glands. A few very inconspicuous filiform tubular spinnerets opening on the margin. Anal aperture about half way between genital aperture and extremity. Diameter averaging 0·50 mm.

Adult male (*fig.* 1) bright yellow, apodema very pale reddish; broad, rather depressed; wing ample and long; genital spike more than half length of body. Ocelli black, lower pair not contiguous (*fig.* 2). Antennæ hairy, ten-jointed; third joint longest; tenth longer than penultimate, tapering to a point, a single knobbed hair at apex and two at side (*fig.* 3). Foot with four digitules (*fig.* 4); tarsus much shorter than tibia. Total length 0·80 mm., including genital spike, which measures 0·30 mm.

Neither eggs nor young larvæ were observed within the galls.

The comparative size of the male and female insects in this species, the female being smaller than the male, is unusual.

Habitat on leaves of *Grewia orientalis*; Punduloya (July).

A species of *Fiorinia* (*F. secreta*) also occurs on this same plant, and forms galls similar to those of the *Aspidiotus*.

EXPLANATION OF PLATE XI.

ASPIDIOTUS OCCULTUS.

(All figures, except No. 5, more or less enlarged.)

Fig. 1. Adult male, dorsal view.
    2. ,, ,, under side of head.
    3. ,, ,, terminal joint of antennæ.
    4. ,, ,, leg.
    5. Leaf, showing galls formed by the insect, nat. size.
    6. Piece of leaf, showing under surface of gall.
    7. Section of gall, showing adult female and puparium.
    8. Male puparium.
    9. Adult female, ventral view.
 10. ,, ,, dorsal view.
 11. ,, ,, extremity of pygidium.

ASPIDIOTUS AURANTII, *Maskell.*

(PLATE XII.)

Aspidiotus aurantii, Maskell, *Trans. N. Z. Instit.*, 1878, p. 109.
Aspidiotus citri, Comstock, *Canadian Entomologist*, Vol. XIII. p. 8.
Aonidiella aurantii, Berlese, *Le Cocciniglie Italiane*, Part III. p. 212.

Female puparium circular; median area convex. Colour pale yellowish grey, semi-transparent, showing the form and colour of the insect beneath it (*fig.* 2). Pellicles central, reddish, obscured by a layer of secretion; a small prominent spot and concentric ring of whitish secretion in centre of first pellicle. Ventral scale well developed, adhering to the insect, as does the dorsal scale also, making the extraction of the insect difficult except by dissolving the puparium in caustic potash. Diameter 2 mm.

Male puparium (*fig.* 3) oblong; dull reddish brown, paler at the margin; pellicle paler, with a raised spot and concentric ring in the centre. Size 1 by 0·60 mm.

Adult female dull orange, ventral surface obscured by the adherent whitish ventral scale (*fig.* 4); flat below, convex above; skin hard and horny; reniform, the abdominal segments retracted and sometimes completely enclosed by the overlapping sides of the thorax (*fig.* 5). Rostrum in living insect generally pushed to one side (*fig.* 4); but after treatment with potash it resumes its normal central position. Pygidium (*fig.* 11) with six well-developed lobes; the mesal and second pairs tricuspid, the third usually bicuspid. Squames large and deeply fringed on their lateral margins; two in each space between the lobes and three beyond; the first squame laterad of second lobe and those beyond the third lobe all deeply cleft at the extremity, forming two branches, of which the outer one only is fringed, and with a short pointed projection between the two branches (*fig.* 12). No circumgenital glands. A number of filiform tubular spinnerets opening on the dorsal surface by oval pores arranged in two diagonal series on each side; a few more opening directly on to the margin near the extremity. Diameter 1 to 1·50 mm.

Adult male (*fig.* 6) orange yellow, apodema dark brown, limbs and antennæ almost colourless. Ocelli purple black. Antennæ hairy, ten-jointed; first two joints short, others elongate, increasing to sixth, then decreasing to tenth; terminal joint (*fig.* 10) with three knobbed hairs. Feet of first and second limbs with four knobbed hairs; hind feet with three knobbed and one simple hair, the latter representing the second ungual digitule; tarsus about two-thirds length of tibia. Genital spike as long as abdomen. Total length 0·80 mm.

Eggs seldom, if ever, seen; the larvæ shedding the eggshell during or immediately after extrusion.

Young larvæ pale yellow; broadly oval; two prominent lobes at posterior extremity, and several marginal papillæ and minute squames.

Habitat in Ceylon on *Agave americana* (American aloe), and on *Citrus decumana* (pomelo), on under surface of leaves. I have not as yet noticed it on orange trees in Ceylon. Reported to be very destructive in orange plantations in America, where it is known as the Red Scale. In New Zealand and Australia it is abundant on orange and lemon trees, and occurs on the fruit as well as on the leaves. Mr. Maskell states that several hundreds of the scales may often be seen on a single fruit in the market. The species occurs at present only sparingly and locally in Ceylon; but, should orange cultivation be attempted on a large scale it would be advisable to be on guard against the increase of this insect.

EXPLANATION OF PLATE XII.

ASPIDIOTUS AURANTII.

(All figures, except No. 1, more or less enlarged.)

Fig. 1. Piece of aloe leaf, with insects *in situ*, nat. size.
2. Female puparium, from above.
3. Male puparium.
4. Female puparium, from below, with insect *in situ*.
5. Adult female, dorsal view.
6. Adult male, dorsal view.
7. Head of male, from below.
8. Foot of male (third pair).
9. „ „ (first pair).
10. Terminal joint of antenna of male.
11. Pygidium of adult female.
12. A single squame from margin of pygidium.

ASPIDIOTUS CAMELLIÆ, *Signoret*.

(PLATE XIII.)

Aspidiotus camelliæ, Sign., *Essai*, 1869, p. 117.
Aspidiotus rapax, Comstock, *Ag. Report*, 1880, p. 307.
Aspidiotus flavescens, Green, *Insect Pests of the Tea Plant*, 1890, p. 21.
Diaspis circulata, Green, 'Catalogue of Coccidæ of Ceylon,' *Ind. Mus. Notes*, Vol. IV. No. 1 (1896).
Nec Kermes Camelliæ, Boisd., *Ent. Hort.*, p. 334.

Female puparium at first round, with central pellicles (*fig*. 3); afterwards usually oblong, with pellicles situated towards the anterior extremity (*figs*. 4 and 6), often distorted by crowding; strongly convex above, the apex sometimes curving forwards until the pellicles assume a nearly vertical position (*fig*. 5); surface more or less roughened by the admixture of small fibres and hairs from the plant upon which it is fixed, the texture consequently varying with that of the plant. Ventral scale well developed, usually entire. Colour greyish ochreous or brownish ochreous, margin whitish, semi-opaque; pellicles castaneous brown, more or less obscured by layers of whitish secretion, usually with a white central boss surrounded by a prominent whitish ring (*figs*. 2, 3, 4); position and form of pellicles best seen from inner side (*fig*. 6). The development of the puparium can be followed by reference to *figs*. 1 to 4 in sequence. Size $1\frac{1}{2}$ to 2 mm. by 1 mm.

Male puparium unknown.

Adult female bright yellow or orange yellow; no indication of eye-spots. Body broadly oval, slightly tapering and pointed behind; abdominal segments distinct before gestation (*figs*. 13 and 14), afterwards the median area becomes more strongly convex and the segments inconspicuous, and there is usually a large shallow depression above the cephalic area (*fig*. 12). Antennæ rudimentary, consisting of a pair of small tubercles, each with a longish curved hair on one side and surmounted by several (four or more) short-pointed processes (*fig*. 17). No parastigmatic glands. Pygidium (*fig*. 16) broad, terminating in a median pair of prominent bluntly tricuspid lobes; other lobes obsolete. On each side are two deep notches with strongly thickened margins, and laterad of each notch is a pointed projection of the margin of the body representing the missing lobes. A pair of deeply and irregularly divided squames caudad of each notch and two or three simple or finely serrated tapering squames extending slightly laterad of the second notch. There is a pair of spines associated with each lobe, a fourth pair about half way up the margin of pygidium and associated with a small marginal indentation. Anal aperture large and conspicuous, very near the extremity; genital aperture near the base of pygidium. Conspicuous thickenings of the body-wall extend inwards from extremity, partially encircling the anal aperture, and upwards on each side beyond the genital aperture. A series of four irregularly linear bodies extends transversely across the base of pygidium. No grouped glands. There are a few tubular spinnerets of the filiform type opening on the dorsal surface by a short series of pores extending from the second marginal notch on each side. Size about 1·12 by 0·75 mm.

Adult male unknown. Prof. Comstock mentions, with respect to his *A. rapax*, that dead and shrivelled males have been observed.

Embryo fully formed within the body of the parent before extrusion.

The young larva (*figs.* 7 and 9) is bright yellow, with colourless eyes. Form broadly oval; abdominal segments crowded together; pygidial segment large and anal aperture conspicuous. Antennæ (*fig.* 9) slender, two longish hairs at extremity and one or two shorter ones on each joint. A pair of prominent serrate lobes at posterior extremity (*fig.* 11). Caudal setæ about three-quarters the length of the body.

Habitat on stems and leaves of tea (*fig.* 15), cinchona, acacia, osbeckia, ivy, and other plants; Punduloya, Newera Eliya, Kandy. Sometimes occurring in such numbers as to completely cover the surface of the parts affected. Found apparently in all parts of the world. A rather serious pest on tea in Ceylon, as it chiefly attacks the young plants of the first and second year, greatly hindering their growth. I have seen a young cinchona tree actually killed by this insect.

The species does not seem to be much infested with hymenopterous parasites.

EXPLANATION OF PLATE XIII.

ASPIDIOTUS CAMELLIÆ.

(All figures, except Nos. 15 and 18, more or less enlarged.)

Figs. 1, 2, 3, and 4. Consecutive stages in development of female puparium.
    5. Female puparium, side view.
    6.   ,,   ,,   from below, with ventral scale removed.
    7. Newly hatched larva, dorsal view.
    8.   ,,   ,,   ,,   ventral view.
    9. Antenna of larva.
  10. Leg   ,,   ,,
  11. Extremity of abdomen of larva.
  12. Adult female, dorsal view.
  13.   ,,   ,,   ,,   ,, before oviposition.
  14.   ,,   ,,   ventral view.
  15. Tea twig, with insects *in situ*, nat. size.
  16. Pygidium of adult female.
  17. Rudimentary antenna of female.
  18. Cinchona branch, with colony of insects, nat. size.

ASPIDIOTUS CYDONIÆ, *Comstock.*

(PLATE XIV.)

Aspidiotus cydoniæ, Comstock, *Agric. Rep.*, 1880, p. 295.

Female puparium at first circular (*fig.* 2), older scales usually more or less oblong; strongly convex (*fig.* 3). Ventral scale thin; remaining attached to the plant. Colour white or pale ochreous. Surface varying considerably with that of the plant to which it is attached (*figs.* 2 and 7). Pellicles centrally situated in early stages, afterwards nearer one extremity; brownish yellow or straw coloured; exposed or slightly obscured by a thin whitish secretion; a central boss on first pellicle. Diameter 1 to 1·50 mm.

Male puparium unknown.

Adult female bright yellow; oval, narrowest behind; convex above, flattish beneath; segments moderately defined (*figs.* 4 and 5). Antennæ rudimentary; a pair of small tubercules each with a longish curved hair at the side. Some (parasitised?) specimens from a cactus plant were of a more elongated form, with the divisions of the segments strongly marked (*fig.* 9). Pygidium (*fig.* 6) broad; median lobes prominent, bluntly tricuspid; lateral lobes obsolete; two deep notches with thickened rims on each side of median lobes; several deeply divided squames caudad of each notch, and one or two simple ones laterad of each notch. Circumgenital glands in four groups, each with from five to seven orifices (Professor Comstock quotes eight to nine for the anterior laterals, and five to seven for the posterior groups, in the type). A series of about eight filiform tubular spinnerets extends diagonally upwards from the second marginal notch, opening by conspicuous oval pores on the dorsal surface; others open on to the extreme margin. Anal aperture large, close to extremity. Genital aperture between the lower lateral spinneret groups. Size averaging 1 mm. by 0·75 mm.

Adult male unknown.

Habitat in Punduloya on the fruit (*fig.* 1), branches and leaves of the edible fig (*Ficus carica*); on fruit of Pomelo (*Citrus decumana*); and on the stems of the tea plant; also on *Cycas*, on *Cactus*, and on the base of the leaves of a small palm; Kandy. In the latter case the scales are roughed and coloured by the incorporation of reddish fibrous matter from the plant (*fig.* 7). Originally described from specimens found on Quince (*Cydonia*) in Florida. Reported also by Mr. Maskell as occurring on orange trees in the Sandwich Islands. This species is evidently very closely allied to *A. Camelliæ*, the principal differences being the presence of grouped glands, and the larger number of tubular spinnerets. The puparia may be distinguished by the paler colour of the pellicles, and by the more delicate nature of the ventral scale; the pellicles are usually more exposed in this species, nor are the central boss and concentric ring so conspicuous.

EXPLANATION OF PLATE XIV.
ASPIDIOTUS CYDONIÆ.
(All figures, except No. 1, more or less enlarged.)

Fig. 1. Piece of fig with insects *in situ*, nat. size.
    2. Female puparium, upper side.
    3.   „   „   side view.
    4. Adult female, dorsal view.
    5.   „   „   ventral view.
    6. Pygidium of female.
    7. Scale of female, from palm plant.
    8. Female puparium, from below.
    9. Parasitised female.

ASPIDIOTUS SECRETUS, Cockerell.

(PLATE XV.)

Aspidiotus secretus, Cockerell, Supplement to *Psyche*, March, 1890, p. 2.

Female puparium concealed between the layers of the dry sheathing petioles of species of bamboo (*figs*. 1, 3). The formation and growth of the puparium is rather difficult to follow. Apparently the young insect takes up its position upon the stem of the plant beneath the sheath. After the completion of the second moult the adult female insinuates itself into and between the layers of the sheath, excavating a cell for itself, and leaving the pellicle exposed upon the surface. When the sheath is removed from the stem, the insects are detached with it, their position indicated by small blister-like swellings upon its inner surface, this view necessarily presenting the ventral surface of the puparium. Pellicles yellow, darkest at margin. Usually the second pellicle only is present, the first becoming detached and falling away during the earlier stages before the second moult. Even when both are present the two pellicles are more or less separated (*fig*. 3). The margin of the second pellicle is minutely pitted with shallow depressions of a polygonal form (*fig*. 5). Diameter of puparium 1·50 to 2·50 mm.

Male puparium (*fig*. 2) oblong, strongly convex above; white, with a yellowish tinge showing through from within. Pellicle brownish yellow, close to anterior extremity. Length 1 mm. The male puparia are situated upon the surface of the stem, sheltered by the sheathing petioles.

Adult female yellow; broadly oval (*fig*. 6); parasitised specimens swollen and more oblong; flattish; segments well defined; skin rather horny. Rostral setæ very long. A small group of parastigmatic glands around the openings of the anterior spiracles. Pygidium (*fig*. 7) broad and triangular, terminating in a large and single prominent median lobe, which is indented on each side; other lobes obsolete or represented only by marginal points. Margin irregularly serratulate, with four deep indentations on each side, from which thickenings of the body-wall extend upwards, marking the position of the obliterated segments of which the pygidium is presumably composed. Numerous finer corrugations between the segmental lines. A small spine on each side of median lobe, and one or two laterad of each marginal indentation. Anal aperture very small, and situated at some distance from extremity. Circumgenital glands in two groups, probably representing the combined upper lateral and lower lateral groups, though there is no trace of any division. Each group with from eighty to ninety orifices. In the typical specimens from Japan these two groups are connected above by a line of single spinnerets, but this connexion is wanting in every specimen that I have examined from Ceylon. I have failed to find any tubular spinnerets, though there are numerous minute inconspicuous pores opening on the dorsal surface. Length 1 to 2 mm.

At the time of the second moult, the insect might readily be mistaken for a species of *Aonidia*, for the adult insect at first shrinks away from its pupal skin, and lies enclosed but completely separate within it (*fig.* 4), being distinctly smaller than during the second stage; but it subsequently increases in size, bursts its covering, and finally grows to many times its former dimensions. I have noticed a similar *Aonidiform* stage in some other species.

Mr. Cockerell informs me that his description (*loc. cit.*) refers to the second stage of the insect. The adult female, being concealed in the substance of the plant, may be easily overlooked.

A rather abnormal species. The form of the male puparium approaches that of a *Diaspis*.

Habitat beneath the sheathing petioles and bracts of a bamboo *Arundinaria sp.* (*fig.* 1). Punduloya (April). Type from Japan, in similar situation on species of bamboo.

In spite of its protected position, this insect is infested by a minute hymenopterous parasite, which proves to be *Homalopoda cristata*, Howard.

EXPLANATION OF PLATE XV.

ASPIDIOTUS SECRETUS.

(All figures, except No. 1, more or less enlarged.)

Fig. 1. Small branch of bamboo, with bracts removed, showing insects *in situ*, nat. size (males on stem, females on inner surface of bract).
2. Male puparium, from above.
3. Female puparium, from below.
4. Second pellicle, containing early adult female.
5. Marginal area of second pellicle.
6. Adult female from below.
7. Pygidium of adult female.

ASPIDIOTUS INUSITATUS, sp. n.

(PLATE XVI.)

Female puparium very large; flattish; at first oval (*figs.* 3 and 4), afterwards greatly elongated posteriorly (*fig.* 2), the supplemental portion usually narrower than the other. Pellicles yellow, always more or less concealed by secretion; approximately central in early adult stage, subsequently becoming eccentric by the backward extension of the scale. Ventral scale as well developed as dorsal, and bearing what appear to be the ventral halves of the pellicles (*fig.* 3). The two surfaces are so much alike that it is often difficult to decide which is the dorsal and which the ventral surface of the puparium after it has been detached from the stem. Colour brownish white or brownish fulvous. Length 3·50 mm. to 7·50 mm. Greatest breadth about 3·50 mm.

Male puparium unknown.

Adult female yellow, pygidium brownish; comparatively large; oblong oval, abdominal area narrowest (*fig.* 5). A small group of glands on the anterior border of each stigmatic opening. Segmental divisions of the abdomen with dorsal and ventral thickened chitinous bands in three series. Lateral margin of body with short transverse chitinous bars (*fig.* 6). Posterior extremity very hard and horny; lobes obsolete; a broad mesal notch, enclosing a minute central point. Deep indentations on each side indicating the boundaries of suppressed segments (*fig.* 5). Margin minutely and irregularly serratulate. No distinct demarcation between the pygidium and the other abdominal segments. Four conspicuous club-shaped processes extend inwards from the margin near the extremity. A pair of hair-like spines mesad of each process; another pair mesad of first lateral indentation; a few others at intervals beyond. No circumgenital glands. The surface of pygidium and horny margin of body closely covered with innumerable minute pores. On the paler lateral margin a raised rim can be distinguished round each pore (*fig.* 6). Anal aperture small; partially surrounded by a thickened rim; situate at a considerable distance from extremity. Length 3 mm.; breadth 1·60 mm.

Adult male unknown.

Habitat on bamboo (*Arundinaria*, sp.), between the stem and the sheathing bracts and petioles. Kelani valley (March).

A very abnormal species, doubtfully placed in this genus on account of its apparent affinity in structure and habits with *A. secretus*, Ckll. An unusually large insect compared with other members of the genus. The elongation of the puparium is probably due to its confined position, where lateral extension would be difficult.

EXPLANATION OF PLATE XVI.
ASPIDIOTUS INUSITATUS.
(All figures, except No. 1, more or less enlarged.)

Fig. 1. Insects on stem of *Arundinaria*, nat. size.
    2. Fully developed puparium.
    3. Female puparium, early adult stage, from below.
    4.   ,,   ,,   ,,   above.
    5. Adult female, from below.
    6. Portions of lateral margin of body, adult female.
    7. Terminal segments of adult female.

AONIDIA, *Targ. Tozz.*

In this genus the development commences as in *Aspidiotus*, but is arrested at the stage already noticed in the description of *Asp. secretus*. The adult female, at the time of the second moult, decreases in size and is entirely enveloped by the second pellicle. Within this receptacle it passes the remainder of its existence and deposits its eggs, or rather young, for all the species at present known are apparently ovoviviparous. The female puparium is principally composed of the second pellicle, which is proportionately large, and its form is dependent upon that of the pellicle, varying from sub-circular to oblong oval, or, as in *A. bullata*, with a distinct posterior extension. The first pellicle is approximately central, and may bear secretionary appendages; in *A. corniger* glassy horns are developed, while *A. bullata* produces a ball of cottony secretion in its larval stage. The ventral scale is complete, consisting of the ventral part of the second pellicle. The overlying secretion covering the dorsal scale is of a very firm structure, resisting the action of liquor potassæ. After maceration it frequently comes away in one complete piece together with the first pellicle, forming the 'superimposed scale' noticed by Signoret.

The male puparium usually resembles superficially that of the female, and is of almost the same size. It consists of secretionary matter, with the single pellicle placed near the centre.

The adult female, which at first nearly fills the cavity of the second pellicle, gradually shrinks until it is less than half the size of the second stage, this shrinkage taking place principally from the posterior extremity. Towards the end of its existence the adult female occupies little more than the cephalic area of the second pellicle. The thoracic segments are much broader than the abdominal. The antennæ are rudimentary; the mouth-parts are usually rather large. The pygidium is very variable in the form of its margin, some species being without true lobes, others having as many as eight. There are no circumgenital glands, and often no tubular spinnerets. The absence of the glands indicates an ovoviviparous habit; the tubular spinnerets have presumably disappeared through disuse, the scale being secreted during the second stage of the insect. The genital aperture is usually close to the base of the pygidium on the ventral surface. The anal aperture is much nearer the extremity on the dorsal surface, and the lining of the rectum is usually thickened and strongly chitinised.

The adult male, in the single species observed, resembles that of *Aspidiotus*.

SYNOPSIS OF SPECIES.

A. Larval pellicle with secretionary appendages.
    (*a*) Scale sub circular; first pellicle bearing glassy horns ... *corniger*.
    (*b*) Scale with distinct posterior extension; larval pellicle with a cottony tuft .... *bullata*.

B. Larval pellicle without secretionary appendages.
    (*a*) Scale cordate; forming shallow pits in stem of plant ... *loranthi*.
    (*b*) Scale oval; not forming pits .... *obscura*.

AONIDIA CORNIGER, *Green.*

(PLATES XVII. AND XVIIA.)

Aonidia corniger, Green, 'Catalogue of Coccidæ,' *Ind. Mus. Notes,* Vol. IV. No. 1 (1896).

Female puparium semicircular, transverse diameter broadest; flattish or slightly convex. Colour light reddish-brown, minutely mottled with paler specks. First pellicle approximately central; either exposed (*fig.* 7) or bearing the horn-shaped processes of the young scale (*fig.* 8). The latter is the normal condition; but, the pellicle being slightly prominent and the processes very brittle, these appendages are frequently rubbed away. The pellicle itself is divided up into three (a median and two lateral) series of distinct plates (*fig.* 5). Second pellicle (*fig.* 19) very large and broad; anterior margin straight; posterior extremity pointed; its dorsal surface concealed by a horny secretion extending slightly beyond its margin (*fig.* 7). The puparium is closed beneath by the ventral parts of the second pellicle. If this ventral scale be carefully dissected off (a rather difficult task, without injury to the insect), the adult female will be seen lying within the hollow of the second pellicle (*fig.* 9). Size of the second pellicle about 1 by 1·25 mm. Size of complete puparium averaging 1·25 by 1·75 mm.

Male puparium oblong; externally very similar in appearance to that of female, but rather smaller and darker in colour; transverse diameter shortest. The single pellicle placed transversely across the scale, near the middle, and usually bearing the larval horn-shaped processes. Colour reddish brown. A broad groove below for the reception of the pupa (*fig.* 15). Size 1·25 by 0·75 mm.

Adult female (*figs.* 10 and 12) much smaller than the second pellicle. Colour pinkish-purple; abdominal segments and flattened margin creamy white. Thoracic area tumescent; greatly extended laterally. Abdominal segments very much contracted, the posterior extremity scarcely projecting below the lateral margins of the thorax. A thin membranous border all round the insect, minutely and irregularly frayed on the abdominal margin. Antennæ rudimentary; three-jointed, with a stoutish bristle on the basal, and a smaller one on the second joint (*fig.* 11). Pygidium (*figs.* 21, 22, and 23) rather short; no lobes, but the margin with large prominent processes which are very variable, irregular, and often unsymmetrical. The variation is endless, no two specimens being exactly alike. In what may be considered as the most normal form there are four principal projections, and eight smaller pointed processes, one of the latter being caudad of each of the principals, and two nearer the base on each side. The margin immediately laterad of the large outer projection is conspicuously thickened and darker-coloured. The median processes are the most subject to variation. There are twelve small spines on the margin of the pygidium, of which six are dorsal and six ventral. No grouped glands or spinnerets of any kind. Near the base, situated inside the body, is a thickened truncate cone, the lower extremity forming the anal aperture (*fig.* 21). The genital aperture is inconspicuous, close to base of pygidium. Length 0·50 mm. Breadth 1 mm.

Adult male (*figs.* 16 and 17) bright reddish; apodema and sides of thorax darker; antennae purplish, except last joint, which is usually pale; legs pale yellowish. Ocelli black; lower pair largest, a prominent colourless tubercle on each side representing the lateral eyes. Antennae hairy; about two-thirds length of body; ten-jointed, first and second short, third longest; others gradually decreasing to tenth, which bears two knobbed hairs, one at apex and one at side (*fig.* 18). Wings ample, broadly rounded at extremity. Foot with four knobbed hair-like digitules; tarsus two-thirds length of tibia; tibia a little shorter than femur. Genital spike half as long as body. Total length 1 mm. Breadth 0·50 mm.

It will be convenient here to trace the growth of the insect from its earliest stage, and the development of the scale. The young are fully formed within the body of the parent. They are comparatively large, and are deposited singly within the cavity of the second pellicle. Young larva (*fig.* 13) broadly oval; antennae long and slender; legs small; caudal setae not visible being very short and folded back upon the surface of the body; colour pale purplish. When ready for the first moult, the larva has become broadly pyriform (*fig.* 20): the dorsal surface is divided up into distinct plates in three longitudinal series eight in each row; each lateral plate is partially subdivided by a thickened line from the margin; the sutures project as ridges on the inner surface (*fig.* 5), and mould the dorsal surface of the early second stage (*fig.* 3). Before moulting, the larva covers itself with a thin whitish scale with thickened radiating lines, and from the centre of the disc a loose white ball of cottony secretion is formed (*fig.* 2). This in its turn is forced up and thrown off by the growth of fifteen long stout solid colourless glassy horn-shaped processes (four of them considerably smaller than the others), which are at first erect (*fig.* 6), but afterwards bend outward from the centre (*fig.* 4). There is besides a stout tapering opaque white process springing from the anal aperture. At this stage the pellicle appears of a deep purple brown colour. It is remarkable that there are apparently no pores, glands, or spinnerets on the larva to account for the secretion of these peculiar processes. The processes themselves are of a dense glassy structure, which is unaffected by either alcohol or benzine, but melts with dry heat, producing a smell as of burning gutta percha. The larval pellicle has a diameter of 0·50 mm. Up till this stage no sexual differences are apparent. The female insect now increases greatly in size, and finally becomes still more broadly pyriform (*fig.* 19). The skin is hard and of a chestnut-brown colour. The dorsal surface is covered with a horny scale, in the centre of which remains the first pellicle with its appendages. The scale has now assumed its greatest dimensions, although the second moult has not been effected. The final change is accomplished by a decrease in size, the adult insect being greatly reduced in bulk, and passing the rest of its existence imprisoned within its former shell.

The male insect, after shedding its first skin, is of a pinkish or creamy white colour. Instead of developing laterally, it becomes slightly elongated, pointed behind, and the segments marked by rounded lobes on the sides. Shortly before assuming the pupal state the ocelli of the future imago begin to appear in the form of dark bands curving round each side of the head (*fig.* 24). These represent both the upper and lower pairs, which in this early stage are seen to be connected. The pupa is rather broad and depressed, the limbs folded along the sides of the body, and all the segmental divisions clearly indicated (*fig.* 35).

Habitat on the upper surface of leaves of *Psychotria thwaitesii* and *Litsea zeylanica*, Punduloya (September, October). The insects are usually ranged along the sides of the midrib and principal veins.

This interesting species is readily distinguished by the remarkable radiating horns of the young scale, a character which has suggested its specific name.

The female insect is attacked by a minute hymenopterous parasite, which has been described by Professor L. O. Howard under the name of *Encarsia aonidiae*.

EXPLANATION OF PLATES XVII. AND XVIA.

AONIDIA CORNIGER.

(All figures, except No. 1, more or less enlarged.)

Fig. 1. Part of leaf, with insects *in situ*, nat. size.
    2. Young scale, dorsal view (tufted stage).
    3.   ,,   female of second stage, dorsal view.
    4.   ,,   scale (horned stage).
    5.   ,,   ,,   ,,   ,,   from below.
    6.   ,,   ,,   (intermediate stage), side view.
    7. Female puparium, dorsal view.
    8.   ,,   ,,   with larval horns persisting.
    9.   ,,   ,,   from below, showing adult female *in situ*.
       (The ventral half of the scale has been dissected off.)
   10. Adult female, dorsal view.
   11.   ,,   ,,   antenna.
   12.   ,,   ,,   ventral view.
   13. Young larva, ventral view.
   14. Head of adult male, from below.
   15. Male puparium, from below.
   16. Adult male, dorsal view.
   17.   ,,   ,,   side view.
   18.   ,,   ,,   terminal joint of antenna.
   19. Second pellicle of female, ventral view.
   20. First pellicle; diagram of dorsum.
21, 22, 23. Pygidium of adult female.
   24. Male larva, shortly before pupating, ventral view.
   25.   ,,   pupa, dorsal view.

AONIDIA BULLATA, sp. nov.
(PLATE XVIII.)

Female puparium (*figs*. 4, 5, and 6) consisting apparently of the second pellicle only, without any supplementary secretional matter. An empty detached scale, seen from below, has exactly the form of a ladle with a short, broad, handle. Anterior portion widest, broadly rounded, very convex, hemispherical, with marginal area slightly flatter; posterior half abruptly narrowed and flattened; the terminal portion curiously and symmetrically sculptured (*fig*. 4). Colour very dark brown — almost black, a large diffused reddish-brown space towards the posterior extremity. Ventral scale very thin and delicate, ruptured on removal of puparium; persisting only as a narrow sub-marginal zone of whitish secretion (*fig*. 5). No trace of the first pellicle can be distinguished, even after prolonged maceration; the paler coloured space near the posterior extremity being slightly depressed and more or less definitely circumscribed, may be mistaken for the missing pellicle, but, upon closer examination, this resemblance is seen to be delusive. It is probable that the larval skin is dislodged during the excessive development of the anterior part of the scale. Posterior margin minutely and regularly indented (*fig*. 9). Length 0·75 mm. Greatest breadth 0·50 mm.

Male puparium (*fig*. 1) irregularly circular, thin and delicate, silvery white; pellicle placed towards anterior margin, rather large, purplish black, bearing on its centre a conspicuous ball of closely felted white cottony secretion. A tangled mass of delicate white curling filaments proceeds from the posterior part of the pellicle and falls on to the scale behind (*fig*. 1). These appendages are easily dislodged, and are consequently absent in many specimens, leaving the black pellicle completely exposed. Under surface with a shallow groove for reception of the insect, bounded by the remnants of the delicate ventral scale (*fig*. 2). Diameter about 1 mm.

Adult female occupying the anterior part of the cavity of the puparium (*fig*. 5); abdominal segments shrunken and withdrawn into the thoracic region; dorsum very convex and rounded to fit the hollow of the puparium; colour reddish purple, abdominal segments yellowish white. I have not been able to examine the living female before gestation, but macerated specimens show an elongated and narrowed abdominal region (*fig*. 7) in correspondence with the form of the puparium. Rostral apparatus comparatively large, very near the anterior extremity. A pair of minute antennal papillae on the anterior margin. The pygidial area is not well defined; there are no grouped glands; the extremity is squarely truncate, bearing eight very small rounded lobes, with two or three minute marginal points in each space (*fig*. 8); there is a slightly thickened short cylindrical tube leading inwards and upwards from the anal aperture. Length of distended specimen 0·60 mm. Breadth 0·45 mm.

Adult male not observed.
Embryo fully formed within body of parent.
Young larva not observed.

The larval skin is very thick and opaque; but there are suggestions of a division into plates as in that of *A. corniger*. The scale at the time of the first moult (*fig.* 11) has a narrow marginal zone of silvery white secretion; there is a central felted ball and posterior mass of curling filaments as in the male puparium.

The female of the early second stage is of the normal oval form; the pygidium showing all the characters of a *Parlatoria* (*fig.* 10), with ten pointed lobes between and beyond which are deeply fimbriated squames; the tubular spinnerets are short and cylindrical, opening by broad lunular pores on to the extreme margin, with the exception of four which open on the dorsal surface near the extremity. The lobes and squames become thickened as the insect develops, and appear on the indurated pellicle which forms the puparium as a series of marginal tooth-like projections (*fig.* 9). It is an interesting fact that the large enveloping second pellicle of *Fiorinia asteliæ*, Mask., from New Zealand has a very similar parlatoria-like margin to the pygidium.

Habitat on leaves of an unidentified tree, Punduloya. The females are ranged along the midrib and principal veins, while the more conspicuous male puparia are scattered irregularly over the surface (*fig.* 3).

A most interesting though very minute and inconspicuous species. The fact of the male puparium being actually larger than that of the female is unusual. The great change in form and character between the earlier and later periods of the second stage of the female is also remarkable. The available material was unfortunately insufficient to allow of this change being properly followed out. The difference of form between the male and female puparia is so great that they might easily be mistaken for distinct species.

EXPLANATION OF PLATE XVIII.

AONIDIA BULLATA.

(All figures, excepting No. 3, more or less enlarged.)

Fig. 1. Male puparium, from above.
    2. ,, ,, from below.
    3. Leaf, with insects *in situ*, nat. size.
    4. Female puparium, from above.
    5. ,, ,, ,, below.
    6. ,, ,, ,, the side.
    7. Adult female, before gestation, ventral view.
    8. Pygidium of adult female.
    9. Posterior extremity of second pellicle, dorsal view.
    10. ,, ,, of female, early second stage, dorsal view.
    11. Scale of young female, at time of first moult.

AONIDIA LORANTHI, sp. nov.
(PLATE XIX. figs. 1-5.)

Females occupying small cavities in the centre of circular swellings on the stems of *Loranthus* (*fig.* 1, *a*). Puparium (*fig.* 2) circular; flat above, convex below. Second pellicle reddish; subcordate, extending almost to the extreme margin of the scale. First pellicle approximately central; greenish or fulvous; with small prominent central boss. The pellicles covered by a very thin transparent yellowish layer of secretion which gives a granulated appearance to the scale. Puparium frequently overlaid by fragments of the cuticle of the plant (*fig.* 2). Ventral scale complete, being the ventral half of the second pellicle; frangible along a definite sub-marginal line which dips sharply downwards in front (*fig.* 3), accounting for the cleft on the median anterior margin of the adult insect. Beyond this line of separation the dorsal and ventral parts of the scale are closely united. Diameter of puparium about 1 mm.

Male puparium not observed.

Adult female semicircular (*fig.* 4); abdominal segments almost completely withdrawn; a deep cleft from the median anterior margin; a broad marginal area flattened and membranous. Colour creamy white, suffused and veined with purple; a large circumscribed area around the rostrum brown and chitinous. Minute antennal papillæ in front of rostrum, each bearing a curved hair. Pygidium (*fig.* 5) rather truncate, with eight small but prominent obscurely tricuspid lobes, of which the median are shortest. Between and beyond the lobes are some very delicate pointed squames. Minute spines at base of lobes. No grouped glands or tubular spinnerets. A transverse series of four irregular linear thickenings of the body-wall across base of pygidium. Length 0·50 to 0·75 mm.

Adult male not observed.

Larvæ not observed.

The bodies of the females examined contained fully developed embryos.

Habitat on stems and branches of *Loranthus, sp.;* Punduloya (January). This is one of those species that form for themselves pits in the substance of the plants upon which they exist. In this particular instance the cavity is really due to a swelling of the tissues immediately surrounding the insect, the bottom of the cavity representing the normal level. Small ring-shaped swellings on the bark mark the former position of insects that have died or become detached.

On the same plant were other scales, distinguished by their more oval form and the absence of the swollen ring, which I at first supposed to be the males of the above species; but upon closer examination they proved to be the mature female puparia of a distinct specie which I have described below under the name of *Aonidia obscura*.

EXPLANATION OF PLATE XIX. (*figs.* 1-5).

AONIDIA LORANTHI.

(*All figures, except No. 1, more or less enlarged.*)

Fig. 1, *a*. Female insects, nat. size, on stem of *Loranthus*.
2. Female puparium, dorsal view.
3. ,, ventral view, with part of ventral scale removed.
4. Adult female, ventral view.
5. Pygidium of adult female.

AONIDIA OBSCURA, sp. nov.
(PLATE XIX. figs. 1, b AND 6-9.)

Female puparium (fig. 6) oval; moderately convex; probably normally covered by a continuous layer of secretion, but in every specimen examined this has been rubbed off, except at the margin, leaving the pellicles exposed. Second pellicle proportionately large; occupying the greater part of the scale; colour bright red-brown; margin regularly crenate (fig. 7). First pellicle paler; approximately central; slightly depressed. Ventral scale complete, being the ventral half of the second pellicle which, in this species, is free to the extreme margin. Greater diameter 0·75 mm.

Male puparium not observed.

Adult female very small, occupying the anterior half of the second pellicle; broadly rounded in front; abdominal segments abruptly narrowed, and with ragged lateral margins; pygidium rather prominent (fig. 8). Colour of dried specimen pale creamy white. Pygidium (fig. 9) with six prominent pointed lobes, the outermost on each side very small; median lobes rather widely separate. Squames narrow, tapering, hairlike; two in the space between median lobes; three very long ones (twice as long as lobes) in each of the two following spaces; and two very small ones beyond the third lobe on each side. Small spines at base of each lobe, and one half way between third lobe and base of pygidium. Diameter of female about 0·50 mm.

Adult male not observed.

Larvæ not observed.

Large, fully formed embryos are visible within the body of the female (fig. 8.)

Habitat on branches of Loranthus; Punduloya (January).

A very small and inconspicuous species, distinguishable from *A. loranthi*, which has the same habitat, by its more oval form and the crenate outline of the large second pellicle.

It appears to be largely subject to the attacks of some hymenopterous parasite, the greater number of the scales examined showing an aperture on the dorsum by which the minute fly has escaped.

EXPLANATION OF PLATE XIX. (figs. 1, b AND 6-9.)
AONIDIA OBSCURA.

(All figures, except No. 1, more or less enlarged.)

Fig. 1, b. Female insects, nat. size
 6. Female puparium, dorsal view.
 7. Second pellicle, outline.
 8. Adult female, ventral view.
 9. Pygidium of adult female.

MYTILASPIS, *Sign.*

Species in which the female puparium is elongated, with the pellicles at the anterior extremity. Ventral scale complete, or divided down the middle; sometimes obsolete.

Male puparium similar to that of female, but smaller and narrower; the single pellicle situated at the anterior extremity. The hinder part of the dorsal scale is often marked off from the rest by a thinner transverse band which acts as a hinge, allowing the extremity of the scale to be raised like a valve at the emergence of the adult male.

Adult female elongate; abdominal segments widest, distinctly divided, often expanded laterally. Circumgenital glands in five groups with comparatively few orifices. Usually a few glandular pores (parastigmatic glands) round the opening of the anterior spiracles. Squames tapering and spine-like. Tubular spinnerets cylindrical, opening on to the margin by prominent oval pores. Anal aperture situate close to base of pygidium; genital aperture near the centre. The female, after oviposition, is greatly reduced in size, occupying the anterior part of the scale and almost completely covered by the pellicles. The supplementary part of the scale acts as an ovisac.

Adult male with the pro- and meso-thorax elongate, throwing the point of insertion of the wings unusually far back. Antennæ with one or more knobbed hairs on the terminal joint.

The several species of this genus are all very much alike, especially in the characters of the pygidium, making their separation extremely difficult. The many small differences that can be plainly *seen*, and appreciated as of specific importance, cannot be so readily *described*. The structure of the ventral scale, whether entire or divided, is of assistance in conjunction with other characters, but cannot be entirely relied upon. The ventral scale is really structurally entire in every case, the apparent division being caused by rupture in removing the scale from its support; but this always occurs along a definite line, and indicates a weakness of the scale at that part which is usually constant in the species. The form of the second pellicle will sometimes assist in the separation of two closely allied species.

I find in Ceylon three distinct species, and a fourth rather distinct form, which is however, at present, ranked only as a variety.

SYNOPSIS OF CEYLON SPECIES.

A. Ventral scale entire; eggs arranged irregularly, or in more than two rows.
    (*a*) Female puparium widened behind, mussel-shaped; ventral scale whitish .... *citricola.*
    (*b*)   ,,   ,,   long and narrow, ventral scale green ...... *cocculi.*
B. Ventral scale divided; eggs arranged in two longitudinal rows.
    (*a*) Female puparium red-brown; division of ventral scale narrow *gloverii.*
    (*b*)   ,,   ,,   pale fulvous; division of ventral scale wide *pallida.*

MYTILASPIS CITRICOLA, *Packard.*

(PLATE XX.)

Aspidiotus citricola (Packard). *Guide to the Study of Insects*, 2nd edition (1870), p. 527.
Mytilaspis citricola (Packard). Comstock, *U.S. Ag. Report*, 1880.

Female puparium elongate, mussel-shell shaped, usually slightly curved, dilated behind, with a narrow flattened marginal area not very sharply defined (*fig.* 2). Colour reddish brown or olivaceous brown, paling on the marginal area. The scale is transversely rugose with numerous curved lines of growth. The pellicles occupy rather more than one-third the total length. First pellicle exposed, pale yellowish. Second pellicle reddish; more or less concealed beneath a layer of secretion, but when separated and cleaned it is seen to have a blunt point on the lateral margin of each of the four basal abdominal segments (*fig.* 8). Ventral scale well developed, whitish; completely enclosing the insect (*fig.* 3); without any median division, except near the anterior extremity, where is an opening for the rostral setæ; not extending to the edges of the dorsal scale, but attached at some little distance from the margin; with a large irregular opening behind, exposing the eggs which are irregularly disposed in the cavity. The surface of the ventral scale is marked and pitted with the impress of the orange rind or leaf upon which it may have been fixed. Length of complete scale 2 to 3 mm. Breadth 0.80 to 1 mm.

Male puparium (*figs.* 5 & 6) smaller, narrower, and of more delicate texture; sides nearly parallel; pellicle occupying nearly one-third of total length. At about one-fifth from posterior extremity is a distinct transverse paler line where the scale is thinner, acting as a hinge, so that the hinder part of the scale can be raised to allow of the exit of the adult insect. Colour varying from pale olivaceous brown to dark reddish brown, the part behind the hinge usually tinged with purplish brown. Pellicle pale straw colour. Ventral scale incomplete, represented by a narrow whitish strip on each side attached a little way within the margin and extending backwards as far as the hinged part of the scale. Size about 1.50 by 0.40 mm.

Adult female creamy white, terminal segment reddish fulvous. Oblong, narrow in front, widest behind. Abdomen deeply segmented laterally. Antennæ consisting of a small tubercle and two stout curved hairs (*fig.* 10). A small group of glands round the anterior stigmatic openings. Pygidium (*fig.* 11) broad, two moderately large median lobes with the free edges crenate and sloping from each side to an obtuse angle, followed closely on each side by a shorter duplex lobe of which the outer lobule is the smaller; other lobes obsolete. There are three marginal indentations on each side beyond the lobes. Squames tapering, spine-like, one pair between the median lobes, a pair in each space between median and second lobes, and a pair springing from each marginal indentation. There are three or four similar spine-like processes on the lateral expansions of the three preceding abdominal segments. Circumgenital glands in five groups: the median represented by a chain of about eight orifices, the upper laterals with twelve to sixteen, and the lower with nine or ten. There are numerous tubular spinnerets; viz., twelve rather large cylindrical ducts opening by conspicuous oval pores on to the margin; and six vertical series of small and inconspicuous capitate ducts opening on the dorsal surface of the pygidium. The margins of the abdominal segments are studded with similar small spinnerets. Anal aperture close to base of pygidium. Genital aperture immediately below the grouped glands. Length 1 to 1·50 mm.

Adult male unobserved. I was unable to rear any perfect males from my Ceylon specimens.

Male larva of second stage (*fig.* 7) similar in form to adult female, but much smaller. Shortly before pupation the ocelli of the imago are indicated by two dark spots on each side of the head, each pair connected by a dark band.

Young larva minute, very active, almost colourless.

Eggs white, disposed irregularly beneath the scale.

There are at least three successive broods of this species in the course of the year. In fact, as is the case with many of our Ceylon scale insects, there seems to be an uninterrupted series of broods.

Habitat on fruit (*fig.* 1) leaves and stems of orange and other species of citrus. When occurring on the leaves, it apparently prefers their upper surface. I have also found it on the leaves of *Toddalia aculeata*, Punduloya.

A very widely distributed species. Found almost wherever any species of *Citrus* is cultivated. I have examples from America, New Zealand, Australia, and Madeira. It is also commonly found on Maltese and other Continental oranges sold in England. I have examined such specimens, and have found the scales to contain living insects and numerous eggs, proving how readily the pest may be introduced into new countries. It is therefore not surprising that this species has become almost world-wide.

Its presence in large numbers upon an orange tree is generally indicated by an unhealthy appearance of the fruit and foliage. A wash of kerosene emulsion (as described in the chapter on insecticides) will be effective in checking the pest if applied when the young larvæ are seen to be on the move. The older insects are too well protected by their scaly covering to be much affected by this treatment.

EXPLANATION OF PLATE XX.
MYTILASPIS CITRICOLA.
(All figures, except No. 1, more or less enlarged.)

Fig. 1. Piece of orange peel, with insects *in situ*, nat. size.
    2. Female puparium, from above.
    3. ,, ,, below.
    4. Adult female, ventral view.
    5. Male puparium, from above.
    6. ,, ,, below.
    7. Second stage of male, shortly before pupating.
    8. Second pellicle of female.
    9. Abdominal extremity of young larva.
    10. Antenna of adult female.
    11. Pygidium of adult female.

MYTILASPIS COCCULI, sp. nov.
(PLATE XXI.)

Female puparium (*fig.* 4) long and narrow, widening very gradually towards the posterior extremity; dark purplish-brown, opaque, with a very narrow irregular paler flattened border; dorsal scale transversely marked by curved lines of growth, the more prominent of which are of a paler colour. Pellicles very pale yellow, transparent, occupying (in the most mature scales) less than one-third the total length—in one instance, less than a quarter; the second pellicle has the abdominal segments without lateral points. Ventral scale greenish, well developed, undivided (*fig.* 5), of rather solid structure, with a small aperture near the anterior extremity, and a large ragged orifice at the hinder end from which the young larvæ make their escape; the surface of the ventral scale is minutely but deeply pitted. It would, however, be dangerous to give specific value to this character, as it is probably dependent upon the surface of the leaf. Length 3 to 4 mm. Greatest breadth 0·75 mm.

Male puparium (*figs.* 2 and 3) similar to that of female, but smaller; lines of growth not so pronounced, a pale transverse band marking the position of the usual hinge near the posterior extremity. Ventral scale greenish, well developed and rather solid, with a broad median longitudinal division. Length 7·50 mm.

Adult female (*fig.* 9) whitish, abdominal extremity reddish-fulvous; cephalic extremity narrowest; abdominal segments widest, with prominent lateral lobes. There is a distinct transverse fold across the anterior part of the head on the level of the rudimentary antennæ; this feature is constant in all specimens examined. The abdominal segments are much wrinkled and seem to be more highly chitinised than the anterior parts, appearing of a darker colour in prepared specimens. The two last abdominal segments each bear four or five stout spines on the lateral margin, and there is sometimes one on the preceding segment also. Parastigmatic glands not readily discernible, but a small group of three or four can sometimes be seen on the border of the anterior stigmatic opening. Pygidium (*fig.* 10) with evenly rounded extremity. Squames and lobes as in *Myt. citricola*, the lobes rather smaller and less prominent; there are occasionally three instead of two squames caudad of the second marginal notch. The marginal pores are large and conspicuous, as are also the cylindrical spinnerets in connexion with them. The dorsal tubular spinnerets are either absent or very inconspicuous. Circumgenital glands in five groups: the median with from one to six orifices, upper laterals with eight to thirteen, and lower laterals with six to eight. Length averaging 1 mm.

Adult male (*fig.* 6) of normal form. Colour creamy white; head, median joints of antennæ, and sides of thorax tinged with pale purple; dorsal plates of mesothorax outlined with reddish; apodema reddish. Antennæ hairy, nearly as long as body; terminal joint with a single knobbed hair at apex (*fig.* 8). Foot (*fig.* 7) with three knobbed hairs—one on claw and two on tarsus; claw long and slender; tarsus as long as tibia. Length about 0·75 mm., of which the genital spike occupies nearly one-fourth.

Eggs white, disposed irregularly beneath the scale (*fig.* 5).

Habitat on the under surface of leaves of *Cocculus macrocarpus*. Kandy (July). The presence of the scale may be detected by a yellowish discolouration of the leaf.

This species, in the structural details of the female insect, is nearly allied to *M. pomorum*, Bouché, which I have seen in Ceylon on the rind of imported Tasmanian apples. The females of that species have the abdominal segments similarly chitinised. *M. cocculi* may be distinguished by the longer and narrow puparium, with its greenish undivided ventral scale. This species also approaches *M. citricola* (Packard), with which I at first classed it; but I now believe it to be distinct. The longer, straighter, and narrower puparium, the frontal fold on the female, and the more highly chitinised abdominal segments, will assist in the determination of the species. From *M. Gloverii* it can be distinguished by the undivided ventral scale and the irregular disposition of the eggs. These various points of difference, though severally small, collectively tend to justify the separation of the species.

EXPLANATION OF PLATE XXI.

MYTILASPIS COCCULI.

All figures, except No. 1, more or less enlarged.

Fig. 1. Part of leaf of *Cocculus*, with insects *in situ*, nat. size.
    2. Male puparium, from below.
    3.   ,,   ,,   from above.
    4. Female puparium, from above.
    5.   ,,   ,,   from below.
    6. Adult male, dorsal view.
    7.   ,,   ,,   foot.
    8.   ,,   ,,   terminal joint of antenna.
    9. Adult female, ventral view.
  10.   ,,   ,,   pygidium.

MYTILASPIS GLOVERII, *Packard*.
(PLATE XXII.)

Coccus gloverii (Packard). *Guide to the Study of Insects* (1869), p. 527.
Aspidiotus gloverii (Packard). *Ibid.*, 2nd edition (1870), p. 527.
Mytilaspis gloverii (Packard). Ashmead, *Orange Insects* (1880), p. 1.

Female puparium (*figs.* 2 and 3) very long and narrow, straight or curved, sides almost parallel, with a well-defined narrow flattened border. Colour variable, brownish-yellow, reddish, or dark reddish-brown, with pale margin; the paler specimens usually occurring on the under surface of the leaf. Pellicles yellow, occupying a little more than one-third of the total length of the scale; the second pellicle elongate (*fig.* 7), with lateral margins of abdominal segments rounded. Ventral scale whitish, well developed, with a median longitudinal division (*fig.* 3) attached a little within the margin. Total length of scale 2·50 to 3 mm.; breadth 0·59 mm.

Male puparium (*figs.* 4 and 5) much smaller and more delicate, long and narrow. Colour pale yellowish-brown, with whitish flattened margin. Pellicle yellow, occupying about one-third the total length. There is the usual hinge-like structure towards the posterior extremity. Ventral scale incomplete, a narrow white strip on each side, attached within the margin. Length 1·50 mm.; breadth 0·50 mm.

Adult female (*fig.* 2) elongate, lateral margins straight and approximately parallel, the abdominal area being only slightly wider than the anterior parts. The meso- and meta-thorax are confluent without any definite line of division, which is, however, indicated by the position of the stigmata. Abdominal segments very short, with lateral margins not much produced; the last three each with two (sometimes three) stout spine-like processes, and numerous minute tubular spinnerets opening on the dorsal surface along the margins and segmental sutures. A very small group (two or three orifices only) of glands in front of the anterior stigmatic openings. After gestation the cuticle of the thoracic parts becomes rather firm and chitinous, and is minutely ribbed transversely. Colour pale purple; extremity of abdomen reddish. Pygidium (*fig.* 8) broad: extremity more truncate and sides more acutely sloped than in *citricola*. Mesal lobes obscurely tricuspid, not conspicuously crenulate, smaller than in *citricola*; second lobes duplex; other lobes obsolete. There are nine pairs of simple tapering squames placed as in *citricola*, but rather longer and stouter. Circumgenital glands in four groups: in my Ceylon examples I find three or four orifices in the median group, six or seven in the upper laterals, and four to six in the lower laterals. Margin with twelve large oval pores in connexion with rather conspicuous tubular ducts. Four series of small capitate ducts opening by large pores on to the dorsal surface.

Anal aperture close to base of pygidium; genital aperture below lower gland groups. Length 1 to 1·25 mm.; breadth less than half a millimetre.

Adult male not observed in Ceylon. Prof. Comstock figures it (from America) as of the normal form, with the parts anterior to the insertion of the wings very long.

Eggs at first white, afterwards tinged with purple; arranged regularly and symmetrically in two rows beneath the scale (*fig.* 3). Although this species is distinctly oviparous, I have noticed well-developed embryos in the bodies of some of the females. Possibly, after depositing a certain number of eggs, a few living young may be directly produced.

Habitat.—I have received this species from Kandy in January, upon leaves and young stems of orange trees.

This species may be distinguished from *citricola* by the narrower and straighter puparium of the female, by the median longitudinal division in the ventral scale, and by the arrangement of the eggs beneath the scale. It is more difficult to point out specific differences in the anatomical structure of the insect itself, but the elongated mesothorax may be of assistance in discriminating between the species. The abdominal segments in *gloverii* bear only two or three marginal spines, while four or five are found in *citricola*. These several differences, though small, appear to be constant.

EXPLANATION OF PLATE XXII.

MYTILASPIS GLOVERII.

(*All figures, except No.* 1, *more or less enlarged.*)

Fig. 1. Orange leaf, with insects *in situ*, nat. size.
    2. Female puparium, from above.
    3.   ,,   ,,   from below.
    4. Male puparium, from above.
    5.   ,,   ,,   from below.
    6. Adult female, ventral view.
    7. Second pellicle of female puparium.
    8. Pygidium of female, ventral view.

MYTILASPIS GLOVERII, *Packard*, var. PALLIDA, *Green*.
(PLATE XXIII.)

Mytilaspis pallida, Green, 'Catalogue of Coccidæ,' *Ind. Mus. Notes*,
Vol. IV. No. I. (1896).

Female puparium (*figs.* 3 and 4) very pale straw colour, with a broad colourless flattened border which extends round the pellicles. Ventral scale broadly divided, showing the eggs disposed symmetrically in two rows. Length 3 to 3·50 mm., of which the pellicles occupy rather less than one-third. Breadth 0·80 mm.

Male puparium (*fig.* 2) with broad flattened margin. The hinge-like structure is not so conspicuous as in the previous species, but the scale posterior to that part is of a darker, more fulvous tint. Length 1·50 mm.

Adult female (*fig.* 8) creamy white, extremity reddish fulvous. The three last abdominal segments each with three or four short stout marginal spines. Abdominal segments rather more chitinised than the anterior parts. I find the number of orifices in the circumgenital glands to be very constant; median with 3 widely separate orifices, upper laterals with 7 or 8, and lower laterals with 4 only. Length 1 to 1·25 mm.

Adult male (*fig.* 5) pale purple; head darker; legs pale yellowish; dorsal plates of mesothorax pale and tinged with reddish. Anterior parts of the body elongated. Terminal joints of antenna with three knobbed hairs (*fig.* 7), one at apex and two at side, the latter very fine. Tarsus shorter than tibia. Foot with three knobbed hairs (*fig.* 6), the one on claw longest. Length 0·62 mm., of which the genital spike occupies one-third.

Habitat on leaves of various unidentified shrubs. Punduloya, Kandy (January).

The difference between this insect and typical *M. gloverii* seems scarcely sufficient to justify its specific separation, but the colour and structure of the puparium—especially the very wide flattened border—incline me to rank it provisionally as a well-marked variety.

EXPLANATION OF PLATE XXIII.
MYTILASPIS GLOVERII, *var.* PALLIDA.
(All figures, except No. 1, more or less enlarged.)

Fig. 1. Leaf, with male and female scales, *in situ*, nat. size.
    2. Male puparium, from above.
    3. Female puparium, from above.
    4.   „   „ from below, showing insect and eggs.
    5. Adult male, dorsal view.
    6. Foot of male.
    7. Terminal joint of antenna of male.
    8. Adult female, ventral view.
    9.   „   pygidium.

DIASPIS, *Costa.*

Species in which the female puparium is more or less circular, with the exuviæ (pellicles) usually placed entirely within the margin, though in some examples the first pellicle may extend beyond the margin of the scale. Ventral scale usually delicate, remaining attached to the plant.

Male puparium small, oblong, opaque white, often tricarinate.

Adult female 'peg-top' shaped, anterior segments widest. Abdominal segments distinct, the divisions marked by transverse rows of conspicuous oval pores communicating with short cylindrical ducts (tubular spinnerets); squames tapering and spine-like; sometimes divided at the tips. There is usually a prominent group of parastigmatic glands round one or both pairs of spiracles. The anal aperture is usually above the level of the genital aperture.

Adult male of normal form. Antennæ with a single knobbed hair at apex.

Only two species of *Diaspis* have at present been found in Ceylon :—

A. Median lobes of pygidium, large and prominent.................. *amygdali.*
B.   „   „   small and sunk into the margin... *jagrææ.*

DIASPIS AMYGDALI, *Tryon.*

(PLATES XXIV. AND XXIVA.)

Diaspis amygdali, Tryon, *Report on Insect and Fungus Pests*, 1889, p. 89.
Diaspis lanatus, Morgan and Cockerell, *Journal of the Institute of Jamaica*, 1892, p. 136.
Diaspis patelliformis, Sasaki, *Bull. Agric. Coll. Imp. Univ. Tokyo, Japan*, Vol. II. p. 107 (1894).

Female puparium (*figs.* 5 to 8) white, yellowish white, or greyish, the colour being principally due to the admixture of particles of the hairs and epidermis of the plant. The female scales are often very inconspicuous, simulating more or less closely the surface upon which they are placed (*fig.* 1). In some cases, however, they remain conspicuously white against the dark green of the plant; I have noticed this more particularly in specimens affecting *Tylophora*. Where the stem of the plant is hairy, the hairs are usually lifted up and incorporated into the substance of the scale, remaining erect in their natural positions (*fig.* 6). The dorsal scale is moderately convex, irregularly circular. Pellicles reddish brown; sometimes exposed (*fig.* 5), sometimes obscured by an overlying deposit of whitish secretion (*fig.* 8); approximately central (*fig.* 5), or placed near the margin (*fig.* 6); from below, the second pellicle is seen as a shallow reddish depression in the white roof of the scale (*fig.* 7). In younger specimens the form of the scale is often oblong with the pellicles situated at one extremity (*fig.* 13); this form is sometimes retained in the mature scales. Ventral scale very thin, remaining attached to the plant. Diameter 2 to 2·50 mm.

Male puparium snowy white; uncarinated (*fig.* 2), or feebly keeled (*fig.* 3). Ventral scale well developed, completely concealing the pupa (*fig.* 4). Pellicle pale straw colour. Length 1 to 1·50 mm. The male scales are usually thickly clustered together, each scale attached, by its anterior extremity, with the hinder parts more or less elevated.

The adult female varies in colour from pale creamy white to pinkish orange, or rosy pink, according to age and food plant (a bright pink variety affects *Callicarpa lanata*); the posterior extremity is always reddish brown; on the under surface a white waxy patch appears on each side of the rostrum, marking the position of the anterior stigmata (*fig.* 11). In old specimens, after oviposition, the under surface of the pygidium is often thickly covered with white waxy secretion (*fig.* 12). The form of the female insect is broadly oval, broadest and rounded in front, narrowing and bluntly pointed behind (*fig.* 10). Segments distinct; the lateral margins prominent and armed with stout spine-like squames. On the dorsal surface of each segment is a pair of large depressed spots (foveæ). Antennæ (*fig.* 19) rather close together; situate immediately in front of the rostrum; each consisting of a stout curved bristle on an irregularly lobed tubercle. Rostrum sunk in a shallow depression. Anterior stigmata surrounded by a well-defined group of parastigmatic glands (*fig.* 22). Pygidium (*fig.* 20) terminating in a pair of prominent pointed lobes, their bases confluent, and their free margins minutely scalloped (*fig.* 21). There is a single very prominent tooth-like lobe laterad of and close to the median lobes, followed at intervals along the margin by three duplex lobe-like marginal prominences. The squames, though apparently tapering and spine-like, are laterally compressed, with the extremities divided into three or more points (*fig.* 21). The squames are perforated at the extremities, and delicate silky filaments can be seen to emanate from them if the insect be removed from its scale and left exposed for a time (*fig.* 11). The number of the squames is somewhat variable, and often differs on opposite sides of the same insect; at the base is a group of six or seven, and there are one or two in the usual positions in each interval between the lobes and marginal prominences. The circumgenital glands are in five groups, with a large but variable number of orifices, the median containing from twelve to twenty-five, the upper laterals thirty to forty-six, and the lower laterals twenty-eight to thirty-eight; the upper laterals in nearly every case containing the greater number. The tubular spinnerets are short and cylindrical, opening by conspicuous oval pores arranged in definite series on the dorsal surface; there are two of these series on each side of the pygidium exterior to the grouped glands, and one on each of he two preceding abdominal segments; there are also similar spinnerets and pores on the margins of the pygidium and preceding four segments. The anal and genital apertures are optically superposed. Length about 1·35 mm. Breadth 1 mm.

Adult male (*fig.* 9) bright red, head darker, legs and terminal joints of antennæ pale. Ocelli black, upper pair larger than lower. Terminal joint of antennæ as long as or longer than ninth, suddenly narrowed near the tip, bearing a stoutish knobbed hair at its apex (*fig.* 26). Foot (*fig.* 25) with three knobbed hairs, one on claw and two on tarsus—the latter often unequal in length; claw long and slender; tarsus considerably shorter than tibia. Genital spike about as long as abdomen. Total length (including genital spike), about 0·75 mm.

The newly hatched larvæ are of two distinct colours, creamy white and bright pink; the difference is possibly sexual. The paler larvæ, after fixing themselves to the plant, secrete from a pair of dorsal pores immediately behind the head a pair of long curling colourless filaments (*figs.* 14 and 15). These are very brittle, and are constantly breaking off, to be replaced by fresh growth from behind, the broken portions accumulating round the body of the insect. I have not noticed these dorsal filaments upon the pink-coloured larvæ. The posterior extremity of the body bears a pair of small but prominent tricuspid lobes (*fig.* 15).

I find a similar difference in the colour of the eggs, some being pale yellow, and others pale pink, the two forms being laid by distinct individuals. It seems probable, therefore, that special females, or females at some special period, produce larvæ of one sex only.

Professors Riley and Howard have published (*Insect Life*, Vol. VI. p. 287) a most careful and valuable paper upon the life-history, habits, and distribution of this insect. By direct observation they ascertained that from eight to nine weeks were occupied in the development of the female insect, and that a constant succession of broods occurred

Habitat.—I have found this species on many different plants in Ceylon, viz., *Callicarpa lanata*, *Tylophora asthmatica*, heliotrope, peach, and several unidentified trees, but it is chiefly noticeable upon cultivated geraniums in gardens, where it is really a most serious pest, the insects being frequently massed upon the stems to such an extent as to kill the plant. The presence of the pest can scarcely escape notice, the stems appearing to be thickly encrusted with a white scurfy matter, which proves to be composed of masses of the male scales. The female scales are much less conspicuous. I find this species at all times of the year, but I think its ravages are worse in the early (dry) months—January, February, and March. Plants that are sheltered from the weather are more particularly subject to attack. This insect is found in widely separate quarters of the globe. It was originally noted by Professor Tryon on peach trees in Australia, and described by him under the name of *Diaspis amygdali*. It subsequently attracted attention in the United States of America, where it attacks a large number of plants, but is especially injurious to peach and plum trees. It was there identified as *Diaspis lanatus* (Morgan and Cockerell); but, as pointed out by Mr. W. M. Maskell (*Trans. N. Z. Instit.*, 1894, p. 44), the American insect is in all specific points identical with the *D. amygdali*, previously described by Professor Tryon. I have myself compared typical specimens from both countries, and can find no good points of distinction. Professor Cockerell records this pest as occurring in Jamaica on grape, peach, cotton, capsicum, and a variety of other plants too numerous to mention. It is said to affect the 'Papaw' tree (*Carica papiya*) in the island of Trinidad. I possess specimens collected by Mr. Albert Koebele in Fiji; and, more recently, what appears to be the same insect has brought itself into notice as an injurious pest on mulberry trees in Japan, where it has been studied by Professor C. Sasaki of the Agricultural College, Tokyo, and again redescribed under the name of *Diaspis patelliformis*. Professor Sasaki states that in Japan the insect breeds only twice in the year.

Its wide distribution and the great variety of plants upon which this insect is able to thrive constitute it a dangerous pest, and it has been fully recognised as such in America, where great efforts are being made to eradicate it. We may consider ourselves fortunate in Ceylon that the insect has chosen such an unimportant plant as the geranium for its head-quarters. It would be desirable, if possible, to exterminate it before it has acquired a more expensive taste. If it were restricted to geranium plants, this would be comparatively easy, by the destruction of all affected plants; but, unfortunately, it has firmly established itself upon a common indigenous shrub—a plant with woolly stems and leaves (*Callicarpa lanata*)—which is too plentiful and widely distributed for any such wholesale measures. From experiments conducted by the U. S. Government entomologist it appears that the scale of this insect is able to resist some of the strongest insecticides; in fact, nothing but pure kerosene emulsion seems to have been really fatal; but, in the case of growing plants, this remedy would be as injurious as the disease, and would prove as fatal to the plants as the parasites. More dilute washes (one of emulsion to nine of water) have proved efficacious at the time the newly hatched larvæ are on the move.

I have bred from this insect a minute hymenopterous parasite, which has been identified by Prof. Howard as *Aspidiophagus citrinus* (Craw.)

EXPLANATION OF PLATES XXIV AND XXIVA.

DIASPIS AMYGDALI.

(All figures, except No. 1, more or less enlarged.)

Fig. 1. Piece of geranium stem, with colony of insects, nat. size.
    2. Male puparium, from above, uncarinated form.
    3. ,, ,, ,, ,, with median carina.
    4. ,, ,, from below.
    5. Female puparium, upper side, from smooth part of stem.
    6. ,, ,, ,, ,, from hairy part of stem.
    7. ,, ,, from below.
    8. ,, ,, from above, with pellicles concealed.
    9. Adult male, dorsal view.
   10. Adult female, dorsal view.
   11. ,, ,, ventral view, before oviposition.
   12. ,, ,, ,, ,, after oviposition.
   13. Female puparium, from *Tylophora* (Chionaspiform variety).
   14. Young larva, side view, with dorsal filaments.
   15. ,, ,, dorsal view.
   16. ,, ,, extremity of abdomen.
   17. ,, ,, foot.
   18. ,, ,, antenna.
   19. Antenna of adult female.
   20. Pygidium of adult female.
   21. ,, median lobes, more enlarged.
   22. Anterior spiracle of adult female.
   23. Adult male, ventral view.
   24. ,, ,, side view.
   25. ,, ,, foot.
   26. ,, ,, terminal joint of antenna.

DIASPIS FAGRÆÆ sp. nov.

(PLATE XXV.)

Female puparium (*figs.* 2, 3, and 6) irregularly circular or slightly oblong, convex. Colour normally opaque white with fine concentric lines of growth, but older specimens become stained with greyish brown. The reddish pellicles are situated either subcentrally (*fig.* 2) or to one side (*fig.* 3)—the latter position being the more frequent; the hinder parts of the first pellicle are obscured by a layer of secretion. Ventral scale represented by a whitish film which remains attached to the plant. Length averaging 2 mm. Breadth 1·25 to 1·75 mm.

Male puparium (*figs.* 4 and 5) oblong; white, with yellowish pellicle at anterior extremity; no trace of carination. Ventral scale very thin and delicate, remaining attached to the plant. Length 1 mm.

Adult female (*fig.* 7) broadly rounded in front and tapering to a blunt point behind; segments well defined, with lateral margins produced into rounded lobes. Colour reddish. Antennæ (*fig.* 8) rather close together, each consisting of a stout, curved hair on an irregular tubercle. Anterior spiracles each surrounded by a large cluster of parastigmatic glands (*fig.* 9), which are indicated in the living insect by white, waxy patches. Pygidium (*fig.* 10) with a pair of oblique, rounded, median lobes sunk into the margin, their bases confluent, their free margins minutely crenulate; followed closely on each side by a small but prominent duplex lobe, and crenulate marginal prominences at intervals. Squames tapering and spine-like, one in each of the first and second spaces, two in each of the third and fourth spaces, and a group of about six at the base. Each of the three preceding segments has a marginal group of similar spine-like squames. The circumgenital glands are in five groups, of which the median contains from eight to ten orifices, the upper laterals from fifteen to eighteen, and the lower laterals from twenty to thirty, the lower groups in every case containing the greater number. The tubular spinnerets are short and cylindrical, a few opening on to the margin, and others opening, by conspicuous oval pores arranged in definite curved series, on the dorsal surface; there are two of these series on each side of the pygidium and one on each of the two preceding abdominal segments, and there are three or four similar pores on each side, close to the circumgenital glands. There are other very delicate filiform ducts communicating with the marginal squames. The genital aperture is between the lower gland groups, and the anal aperture slightly anterior. Length 1 to 1·10 mm. Breadth 0·80 to 1 mm.

Adult male of normal form and reddish colour. Terminal joint of antennæ equal to 9th, and bearing a single knobbed hair at apex (*fig.* 12). Foot (*fig.* 11) with three knobbed hairs (one on claw and two on tarsus); claw proportionately shorter and stouter than in *D. amygdali;* tarsus considerably shorter than tibia. Genital spike about as long as abdomen. Length (including genital spike) 0·85 mm.

Habitat.—I found this insect thickly clustered on the buds of *Fagræa ceylanica* in the Haldumulla district. The winged males were emerging in April.

This species is very close to *D. rosæ*, Sandb. The uncarinated scale of the male will help to distinguish the Ceylon insect.

EXPLANATION OF PLATE.

DIASPIS FAGRÆÆ.

(All figures, except No. 1, more or less enlarged.)

Fig. 1. Flower buds of *Fagræa*, with scales *in situ*.
    2. Female puparium, from above (central pellicles).
    3.   ,,   ,,   ,,   ,, (lateral pellicles).
    4. Male puparium, from above.
    5.   ,,   ,,   from below.
    6. Female puparium from below.
    7. Adult female, ventral view.
    8.   ,,   ,,   antennæ.
    9.   ,,   ,,   one of the anterior stigmata.
   10.   ,,   ,,   pygidium.
   11. Foot of adult male.
   12. Terminal joint of male antenna.

FIORINIA, *Targ. Tozz.*

In this genus the adult female is smaller than, and is completely enclosed in, the second pellicle, which is proportionately large. The female at the second moult shrinks in size, and thus becomes detached from the skin of the previous stage. The eggs or young larvæ are deposited within this receptacle. The female puparium, consisting principally of this enlarged second pellicle, with sometimes a slight secretional margin, is elongate, with the first pellicle projecting beyond the anterior margin. It bears the same relation to *Chionaspis* or *Diaspis* that *Aonidia* does to *Aspidiotus*. The male puparium is elongate, opaque white, with or without carinæ, the pellicle placed at the anterior extremity. The scales are usually exposed, but in one Ceylon species (*scrobicularum*) they are concealed within the scrobiculæ (glandular pits) of the leaf, while in another species (*secreta*) the female puparia are enclosed in small galls. The pygidium of the adult female usually bears well-developed median lobes, with sometimes two smaller ones on each side. There are normally five groups of circumgenital glands, of which the median and upper lateral groups sometimes coalesce to form an arch, but in the gall-making species the grouped glands are entirely absent.

I find five species of *Fiorinia* in Ceylon. Of these, the first three (*fioriniæ*, *saprosmæ*, and *similis*) are very closely allied, so much so as to be superficially almost unseparable; but the minute structural details of the adult female yield good specific characters. The structure of the rudimentary antennæ will be found useful for the purpose of synopsis.

SYNOPSIS OF CEYLON SPECIES OF FIORINIA.

A. Female puparium exposed
    (*a*) Antennæ of female with a long conical spike .... *fioriniæ*.
    (*b*) Antennæ of female without conical spike.
        (1) Antennæ close together, with a median tricuspid tubercle .... *saprosmæ*.
        (2) Antennæ widely separate, without median tubercle .... *similis*.
B. Female puparium concealed
    (*a*) In glandular pits (scrobiculæ) .... *scrobicularum*
    (*b*) In small rounded galls on surface of leaf .... *secreta*.

FIORINIA FIORINIÆ, *Targ. Tozz.*

(PLATE XXVI.)

Diaspis fiorinia, Targ. Tozz. (1867), *Studii Cocciniglie*, p. 14.
Fiorinia pellucida, Targ. Tozz., *Catal.* (1868), p. 436.
Chermes areccæ, Boisduval, *Insect. Agric.* (1868).
Fiorinia camelliæ, Comstock, *Agric. Report*, 1880, p. 329.
Uhleria camelliæ, Comstock, *Second Rep. Depart. Entom., Cornell Univ. Experiment Station*, 1883, p. 111.
Fiorinia palmæ, Green, 'Catalogue of Coccidæ of Ceylon,' *Ind. Mus. Notes*, Vol. IV. No. 1 (1896).

Female puparium (*figs.* 2 and 7) composed almost entirely of the oblong second pellicle, covered with a thin transparent layer of colourless secretion, which extends very slightly beyond its margin. The first pellicle is very small, pale straw colour, and situated at the anterior extremity, projecting beyond the margin. Colour of second pellicle dull reddish fulvous; median area darker, sometimes forming a well-defined longitudinal stripe (*fig.* 7), which is intensified by the presence of a more or less prominent median ridge. The posterior margin of the second pellicle (*fig.* 8) is similar to that of the pygidium of the adult female. The puparium is closed beneath by a transparent membrane, which is the ventral portion of the second pellicle. The adult insect occupies the anterior portion of the scale, the hinder portion being packed with the pale yellow eggs arranged in two rows. Length about 1·25 mm. Breadth 0·50 mm.

I have not observed the male puparium of this species. Signoret describes it as being similar in structure and appearance to that of the female, but smaller and narrower. This must surely be a mistake. In all the allied species the male scales are snowy white, and usually tricarinate as in *Chionaspis*.

Adult female (*fig.* 3) flattish, marginal area especially thin; abdominal segments (after oviposition) very much contracted, their lateral margins overlapping the base of the pygidium. Signoret describes the insect as very elongated, three times as long as broad. This must refer to the early adult stage, when it would almost fill the puparium. Colour pale yellow; median abdominal area reddish brown. In the living insect minute black eyes are visible close to the margin. The rudimentary antennæ (*fig.* 5), situated in a median depression on the anterior margin, are apparently two-jointed, the basal joint consisting of a broad fleshy tubercle bearing a stout curved bristle on one side, the second joint taking the form of a stout pointed spine, which in some specimens bears a small lateral point near the extremity (*fig.* 6). The margins of the abdominal segments bear a few spinous tubercles (*fig.* 4). Pygidium (*fig.* 4) deltoid; apex rather truncate, with a distinct median notch, the sides of he notch occupied by the oblique median lobes, which are finely serrated along their free edge; two smaller prominent lobes on each side near the extremity, followed by several indentations and minute serrations. Three, or sometimes four, rather conspicuous tubular spinnerets open by cylindrical ducts on to the margin of the pygidium. The circumgenital glands are arranged in a more or less complete arch, but the lower lateral groups can usually be separated. The arch comprising the median and upper lateral groups contains from 25 to 30 orifices; the lower lateral groups have each from 12 to 16 orifices. Size (after oviposition) about 0·75 × 0·50 mm.

Adult male not observed.

Habitat on under surface of leaves of the camellia plant and the cocoanut palm. Punduloya (September). I have not yet found this insect upon tea, but its presence upon such a closely allied plant as the camellia makes it extremely probable that our staple product may be attacked by this pest. Its presence in large numbers upon cocoanut leaves must be injurious, the affected leaves turning a sickly yellow colour. The same insect is said to affect the Areca palm. The species is widely distributed, occurring in Europe, America, and Australia.

EXPLANATION OF PLATE XXVI.

FIORINIA FIORINIÆ.

(All figures, except No. 1, more or less enlarged.

Fig. 1. Piece of cocoanut leaf, with insects *in situ*, nat. size.
    2. Female scale, from above.
    3. Adult female, ventral view.
    4. ,, pygidium.
    5. ,, antennæ.
    6. ,, antenna, branched form.
    7. Female scale, from camellia plant.
    8. Extremity of second pellicle.

FIORINIA SAPROSMAE, Green.

(PLATE XVII.)

Fiorinia saprosmae, Green, 'Catal. of Coccidæ,' *Ind. Mus. Notes*, Vol. IV. No. 1 (1896).

Female puparium (*fig.* 2) very similar to that of *F. fioriniæ*, but rather larger; oblong oval, a very narrow margin of colourless transparent secretion surrounding the large second pellicle. Colour clear orange yellow, through which the form of the adult insect and the ova can be distinguished; towards the hinder extremity is generally a paler region caused by the accumulation of the white eggskins beneath the scale. There is a median logitudinal ridge, which is often of a darker (reddish) colour. Before oviposition the adult insect occupies the posterior portion of the puparium (*fig.* 5), but as the eggs are deposited the insect is shifted forwards until it lies in the anterior half (*fig.* 3), the eggs being arranged in a double row behind. Length 1·75 to 2 mm. Breadth 0·75 mm.

Male puparium (*fig.* 10) opaque white; obscurely tricarinate; pellicle pale yellow. The puparia are usually collected together in small groups, partially concealed beneath a quantity of white filamentous and flocculent matter secreted by the larvæ. Length 1·12 mm. Breadth about 0·40 mm.

Adult female (*fig.* 4) pale yellow, pygidium brownish; margin flattened and almost colourless; no eyespots. Before oviposition the insect is broadly oval (*fig.* 3); afterwards it becomes almost semicircular in outline (*fig.* 4). The abdominal segments are completely retracted, forming a concavity from which the pygidium projects. Just within the anterior margin are the rudimentary antennæ (*fig.* 8), each consisting of two very short joints, with a longish stout bristle springing from the basal one; a small tricuspid tubercle separates the two antennæ. On the lateral margins of the body are several groups of prominent jointed (?) tubercles (*fig.* 9), which may perhaps be used by the insect for shifting its position within the puparium. Pygidium (*fig.* 7) rather broad and bluntly triangular; extremity with a deep median indentation occupied by the median lobes, which are united at the base by a curved transverse piece; the lobes are very slightly serrate at the extremity. Lateral lobes obsolete. Margin of pygidium irregularly sinuate, with several small prominent points and indentations. Tubular spinnerets small and inconspicuous, filiform. Circumgenital glands in five groups, the median and upper laterals forming together a more or less complete arch; lower laterals with eighteen or nineteen orifices, upper laterals with fifteen to twenty, median with six to eight arranged in an irregular chain. Length (after oviposition) 0·75. Breadth 0·62 mm.

Adult male (*fig.* 11) minute; bright orange yellow. Ocelli black; upper pair separate by about their own diameter, lower pair by half their diameter. Antennæ normal, ten-jointed, hairy; tenth joint with a knobbed hair at apex (*fig.* 12). Foot (*fig.* 13) with three digitules, one on claw and two on tarsus; claw rather long; tarsus shorter than tibia.

Female of second stage (*fig.* 6) elongated; the two abdominal segments immediately preceding the pygidium produced laterally into longish pointed processes.

Habitat on leaves of *Saprosma zeylanicum*, Punduloya. The female scales occur on both surfaces of the leaves. The male scales are found only on the under surface. The male colonies are rather conspicuous, owing to the white flocculent matter covering them.

This species may be distinguished from *F. fioriniæ* by the absence of lateral lobes on the pygidium; by the less conspicuous tubular spinnerets; by the structure of the antennæ in the female, and the presence between them of a prominent chitinous tubercle. The marginal tubercles in *saprosmæ* are more numerous than in *fioriniæ*, and more truncate.

A minute hymenopterous parasite keeps this species in partial check. It has been recognised by Professor L. O. Howard as *Arrhenophagus chionaspidis*, Auriv.

EXPLANATION OF PLATE XXVII.

FIORINIA SAPROSMÆ.

(All figures, except No. 1, more or less enlarged.)

Fig. 1. Leaf of saprosma zeylanicum, with insects, nat. size.
    2. Female puparium, from above.
    3. ,, ,, from below, showing insect and eggs.
    4. Adult female, ventral view.
    5. Female puparium, showing position of insect before oviposition.
    6. Female of second stage.
    7. Pygidium of adult female.
    8. Antennæ of female.
    9. Marginal tubercles.
    10. Male puparium, from above.
    11. Adult male, dorsal view.
    12. ,, ,, terminal joints of antenna.
    13. ,, ,, foot.

FIORINIA SIMILIS, sp. nov.
(Plate XXVIII.)

Female puparium (*fig.* 6) very similar to that of *F. saprosma*, but more delicate and transparent, and without any dark median stripe. There is an indistinct median longitudinal ridge, with a short lateral ridge on each side near the hinder extremity. Colour pale clear yellow or orange, transparent, the form of the insect and eggs plainly visible through the scale (*fig.* 6). Secretional margin very narrow. On the under surface there are two waved lines of opaque white secretion extending from the anterior margin nearly to the middle of the scale (*fig.* 11) which are not found in *saprosma*. Length, 1·75 to 1·90 mm. Breadth 0·75 mm.

Male puparium (*fig.* 2) white, tricarinate, the ridges prominent and powdery. Puparia not covered with flocculent matter as in *saprosma*, but the surface of the leaf surrounding the male colonies thinly coated with a fine whitish powder (*fig.* 1). Length 1 mm. Breadth about 0·40 mm.

Adult female (*fig.* 4) pale yellow; margin flattened and colourless; pygidium brownish; anterior extremity not so broadly rounded as in *saprosma*; abdominal segments contracted till the extremity of the pygidium is almost level with their lateral margins. Several groups of truncate marginal tubercles (*fig.* 7). Antennæ (*fig.* 5) wider apart, and with no median tubercle. Rostrum (*fig.* 12) conspicuous on account of the large, blackish, recurved anterior processes. Pygidium (*fig.* 8) deltoid, narrower and more pointed than in *saprosma*. Median lobes very small, erect, and separate; no deep median notch; lateral lobes obsolete. Circumgenital glands in five groups; upper and lower laterals, each with from thirteen to eighteen orifices; median group consisting of three or four separate orifices. Tubular spinnerets very minute and inconspicuous. Length (after oviposition) 0·75 mm. Breadth 0·50 mm.

Adult male reddish orange. Antennæ with a knobbed hair at extremity. First and second pairs of feet with four digitules (*fig.* 9), hind feet with three digitules (*fig.* 10).

Young larva yellow, oval, tapering behind. Abdominal extremity (*fig.* 3) with several tubercles and fleshy spines. Eggs pale yellow.

Habitat on leaves of unidentified shrub; Punduloya. The females occur on both surfaces of the leaves, the male scales on the under surface only.

This species can be distinguished from *saprosma* by the more widely separate antennæ without median tubercle (compare *Pl.* XXVIII. *fig.* 5, with *Pl.* XXVII. *fig.* 8), also by the narrower and more pointed pygidium with its small erect median lobes. The adult males can be separated by the number of digitules on the anterior feet.

EXPLANATION OF PLATE XXVIII.
FIORINIA SIMILIS.
(All figures, except No. 1, more or less enlarged.)

Fig. 1. Leaf, with male and female scales *in situ*, nat. size.
 2. Male puparium, from above.
 3. Abdominal extremity of young larva.
 4. Adult female, dorsal view.
 5.   ,,   ,,   antennæ.
 6. Female puparium, from above.
 7. Marginal tubercles of adult female.
 8. Pygidium of adult female.
 9 Anterior foot of adult male.
 10. Hind foot of   ,,   ,,
 11. Female puparium, from below, showing position of insect before oviposition.
 12. Rostral apparatus of adult female.

FIORINIA SCROBICULARUM, *Green.*

(PLATE XXIX.)

Fiorinia scrobicularum, Green, 'Catal. of Coccidæ,' *Indian Mus. Notes,* Vol. IV. No. 1 (1896).

Female puparium (*figs.* 4 and 10) elongate, narrow in front, broadening behind; posterior half rather abruptly depressed, the large second pellicle covered by a very thin layer of secretion which scarcely projects beyond the margin. Colour pale yellow, median area tinged with reddish, the adult female indistinctly visible through the scale. First pellicle long and narrow. Length 1 mm. Greatest breadth 0·30 mm.

Male puparium (*fig.* 2) opaque white with the pellicle pale yellow, elongate, narrow, uncarinated, hinder part flattened, surrounded by some loose white flocculent secretion which projects from the mouth of the cavity in which the scale is concealed (*fig.* 3). Length 1·15 mm.

Adult female (*figs.* 5, 6, and 7) elongate, narrower in front, abdominal segments widest; division of segments rather indistinct. Body rather thick dorso-laterally, the tumescence ending abruptly at the base of the pygidium, which is thin and deflexed (*fig.* 7). Small blackish eyespots are indistinctly visible on the anterior dorsal margin in living examples. Antennæ (*fig.* 8) of two short joints with a stout bristle on the side of the basal joint. No marginal tubercles. Pygidium (*fig.* 9) triangular, extremity with a deep median notch, on the sides of which are the small oblique median lobes, followed closely by two small prominent marginal points representing the lateral lobes. There are three small indentations on the lateral margins, and three or four small spines. Circumgenital glands in five distinct groups, the median with four to six orifices, upper laterals with seven to eleven, and lower laterals with seven to thirteen. Tubular spinnerets cylindrical, rather inconspicuous, three on each side, opening on to the margin. Anal aperture considerably anterior to the genital opening, almost on a level with the median group of glands. Length 0·60 mm.

Adult male not observed.

Eggs pale yellow, comparatively large, almost a quarter the size of the parent insect.

Habitat:—In the glandular pockets (scrobiculæ) at the angles of the veins on the under surface of the leaves of *Gærtnera Koenigii,* Punduloya (July). The female scales are completely concealed, but their presence may be detected by a yellowish discolouration on the upper surface of the leaf immediately above the affected part. The male scale is only partially concealed. The small white spots at the angles of the veins of the leaf (*fig.* 1) mark the position of the male insects. A very small and inconspicuous species, unlike any of its congeners. The relatively large size of the male scale, which is longer than that of the female, is unusual.

EXPLANATION OF PLATE XXIX.

FIORINIA SCROBICULARUM.

(All figures, except No. 1, more or less enlarged.)

Fig. 1. Insects *in situ*, on leaf of Gærtnera.
    2. Part of leaf, with scrobicula and male puparium.
    3. Male puparium, removed from scrobicula.
    4. Female puparium, from above.
    5. Adult female, ventral view.
    6.   ,,   dorsal view.
    7.   ,,   side view.
    8.   ,,   antennæ.
    9.   ,,   pygidium.
    10. Part of leaf, with scrobicula opened, showing female puparium in position.

FLORINIA SECRETA, Green.

(PLATE XXX.)

Fiorinia secreta, Green, 'Catal. of Coccidæ,' *Ind. Mus. Notes*, Vol. IV. No. 1 (1896.)

Female puparium consisting almost entirely of the inflated second pellicle enclosed in a minute rounded gall (*figs.* 1 and 5). The scale is of a peculiar form (*fig.* 6), pear-shaped, with the ventral and cephalic areas strongly convex, and with an oblong circumscribed depression on what must be considered the dorsal surface. The insect in this stage curiously resembles the pupa of a Syrphid fly. Colour pale yellow, the form and colour of the contained insect partially visible through the covering. Some white waxy matter usually adheres to the lower ventral surface of the scale, and the cavity of the gall is imperfectly lined with a similar material. The puparium rests in an erect position within the gall (*fig.* 5), the posterior extremity always directed towards the opening on the under surface of the leaf, which is normally closed by the small ochreous first pellicle (*fig.* 4); but this latter is frequently displaced or missing, possibly dislodged by the escaping larvæ. The margin of the posterior extremity is closely fringed with fine sharply pointed thorn-like processes (*fig.* 14). Length 0·90 mm. Greatest transverse diameter 0·50 mm. The gall measures about 1 mm. by 0·75 mm.

Male puparium opaque white; slightly concave above, uncarinated; slipper-shaped both in outline (*fig.* 2) and longitudinal section (*fig.* 3). Anterior half covered by the bright yellow pellicle, which is transversely wrinkled (*fig.* 2). A section through the leaf at the point occupied by the puparium shows it to be sunk in a depression in the substance of the leaf (*fig.* 3), a corresponding prominence appearing on the other side. Length 0·75 mm. Breadth 0·25 mm.

Adult female (*figs.* 7 and 8) bright yellow, with inconspicuous blackish eye-spots on the anterior margin. A median longitudinal section through the pear-shaped second pellicle will show the adult female resting loosely inside in an erect position (*fig.* 13). Thoracic segments broadly rounded and strongly convex dorsally; abdominal segments suddenly narrowed and tapering to a point. Antennæ minute, rudimentary, situate on the anterior margin, two-jointed, terminal joint a small conical tubercle, basal joint a flattish oval plate with a stout curved hair on one side (*fig.* 9). Spiracles elevated on small round tubercles (*fig.* 8). Pygidium long and narrow, terminating in a conspicuous prominent narrow median process, which is dilated and indented at the apex (*fig.* 10); the lateral margin near the extremity is very irregularly and variously crenulate or dentate, the two sides being often asymmetrical (*figs.* 10, 11, 12). There are no true lobes, squames, or spires; no circumgenital glands, and no tubular spinnerets. The surface of the pygidium is longitudinally marked with numerous fine lines and wrinkles. Genital aperture close to base of pygidium;

anal aperture a little nearer the extremity. Length 0·75 mm. Greatest breadth 0·50 mm.

Adult male not observed.

No eggs or young larvæ were found in the puparia or galls. From the absence of circumgenital glands in the adult female it is probable that the insect is ovoviviparous. The young larvæ would readily make their escape through the aperture at the base of the gall.

Habitat on leaves of *Grewia orientalis*. Pundaloya (June, July). The female insects occupy minute rounded or conical galls on the upper surface of the leaf; the male scales are sunk in shallow depressions on the under surface.

The development of this insect is essentially that of a *Fiorinia*, though the form of the female puparium is very abnormal and totally unlike that of any of its congeners. It is interesting to find on this same plant three galls, outwardly almost identical, one formed by an *Aspidiotus* (*occultus*), a second by the above-described *Fiorinia*, and a third produced by the action of a microscopic fungus. It would seem that the plant must be predisposed to the formation of galls under the stimulus of irritation. I notice a curious difference between the galls of the *Aspidiotus* and the *Fiorinia* after desiccation. The former, in drying, become of a darker tint than the rest of the leaf; while the exact opposite occurs in the other case, the dried *Fiorinia* galls being considerably paler than their surroundings. No such distinction is apparent in the fresh galls.

EXPLANATION OF PLATE XXX.

FIORINIA SECRETA.

(All figures, except No. 1, more or less enlarged.)

Fig. 1. Leaf of *Grewia orientalis*, with galls, nat. size.
  2. Male puparium, *in situ*.
  3. Section of leaf showing male puparium from the side.
  4. Under surface of leaf, showing opening of female gall closed by the first pellicle.
  5. Section of gall showing female puparium.
  6. Female puparium, removed from gall.
  7. Adult female, side view.
  8.   „   ventral view.
  9.   „   antenna.
  10. „   pygidium.
  11, 12. „ variation in form of extremity of pygidium.
  13. Section (diagrammatic) through puparium, showing position of adult female.
  14. Extremity of second pellicle, dorsal view.

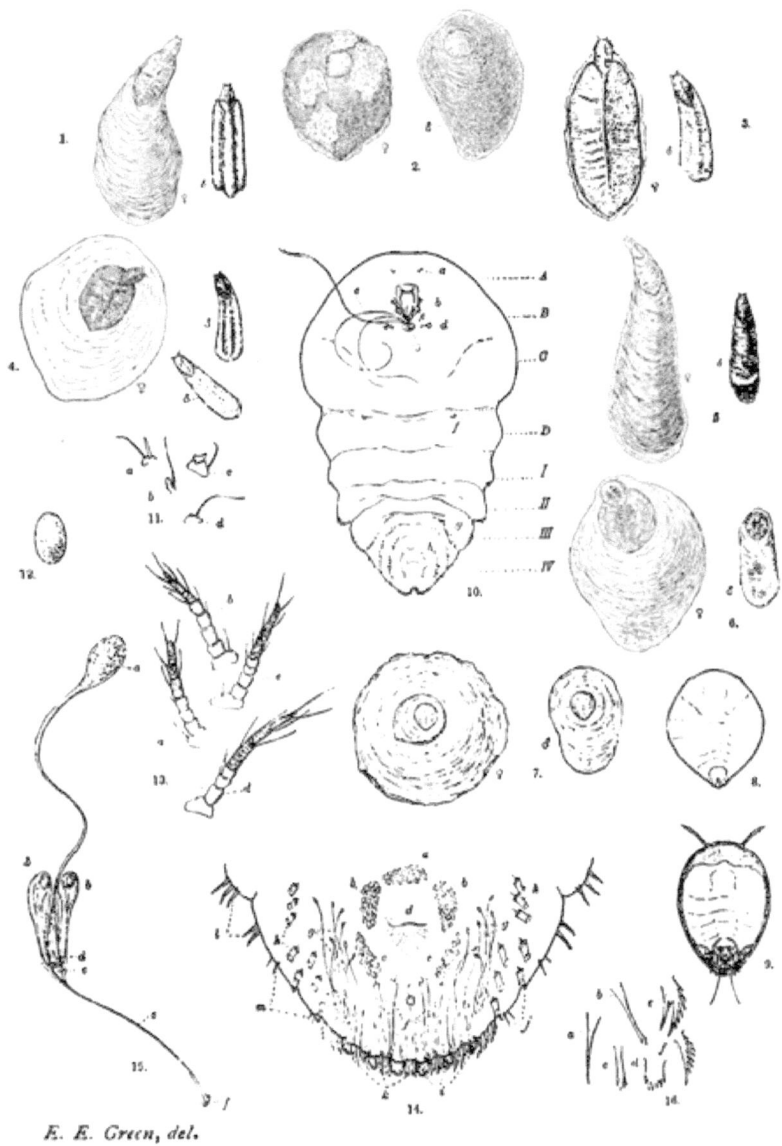

E. E. Green, del.

STRUCTURAL CHARACTERS OF FEMALE AND LARVAL DIASPINÆ.

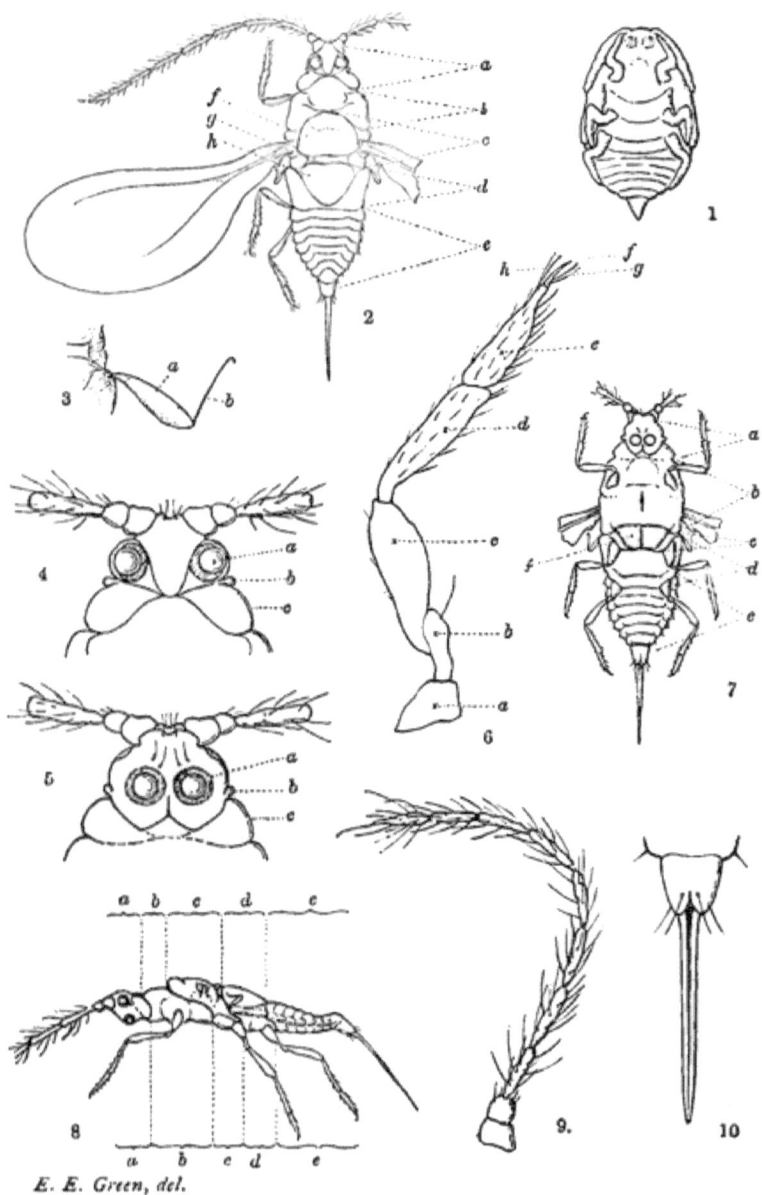

STRUCTURAL CHARACTERS OF MALE DIASPINÆ.

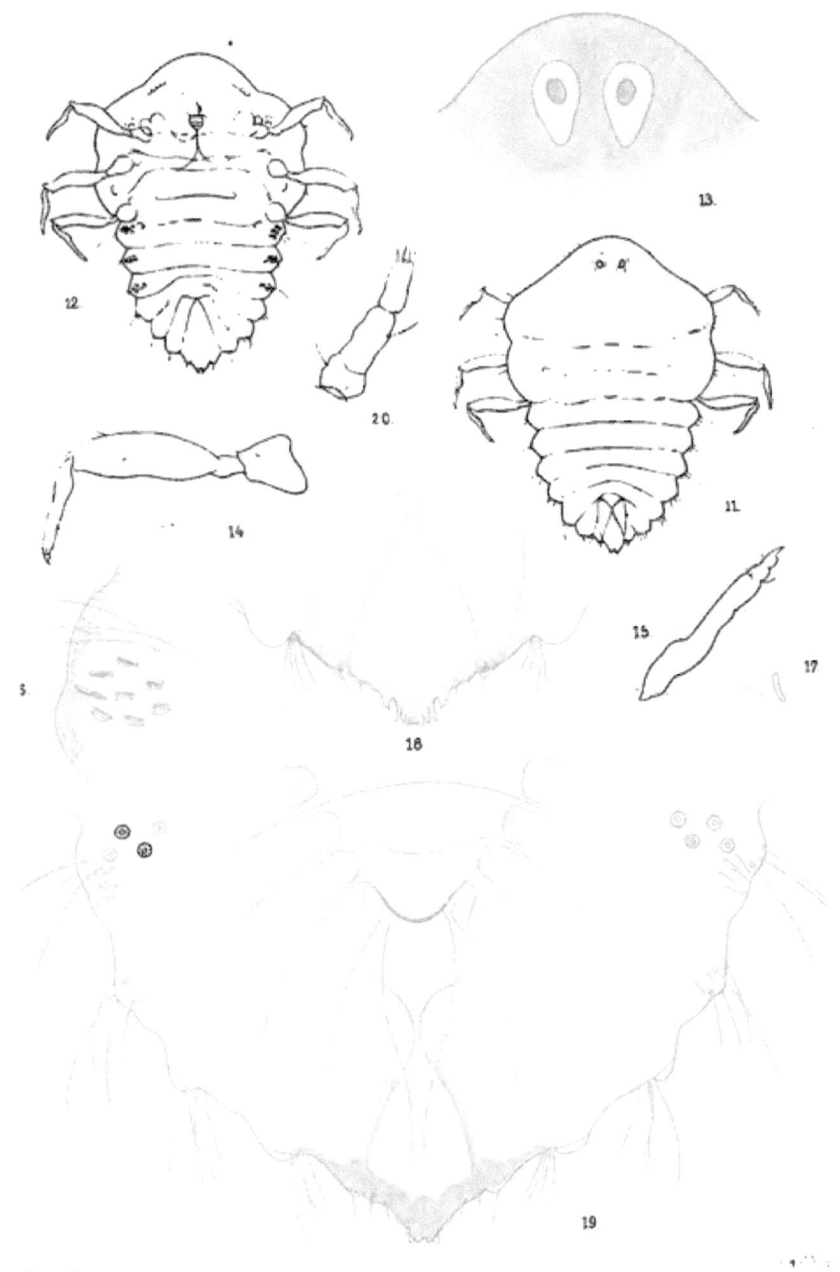

CONCHASPIS SOCIALIS.

ASPIDIOTUS TRILOBITIFORMIS.

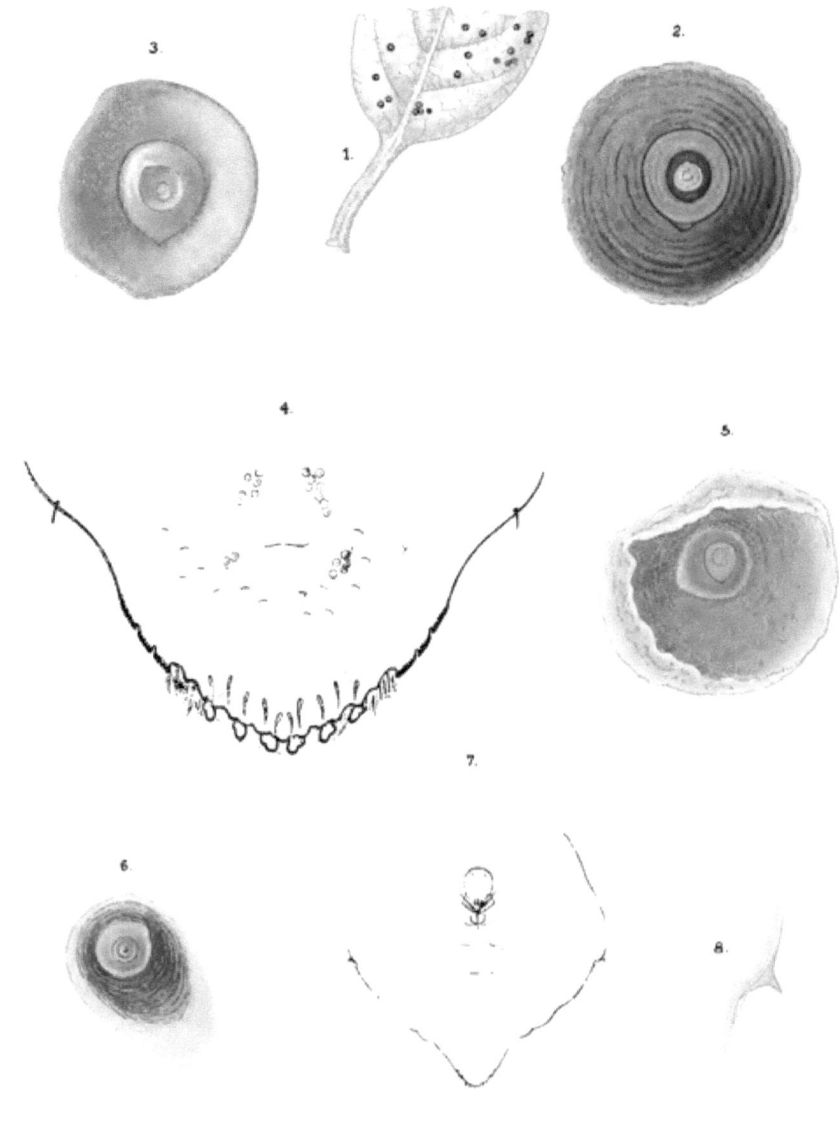

ASPIDIOTUS FICUS.

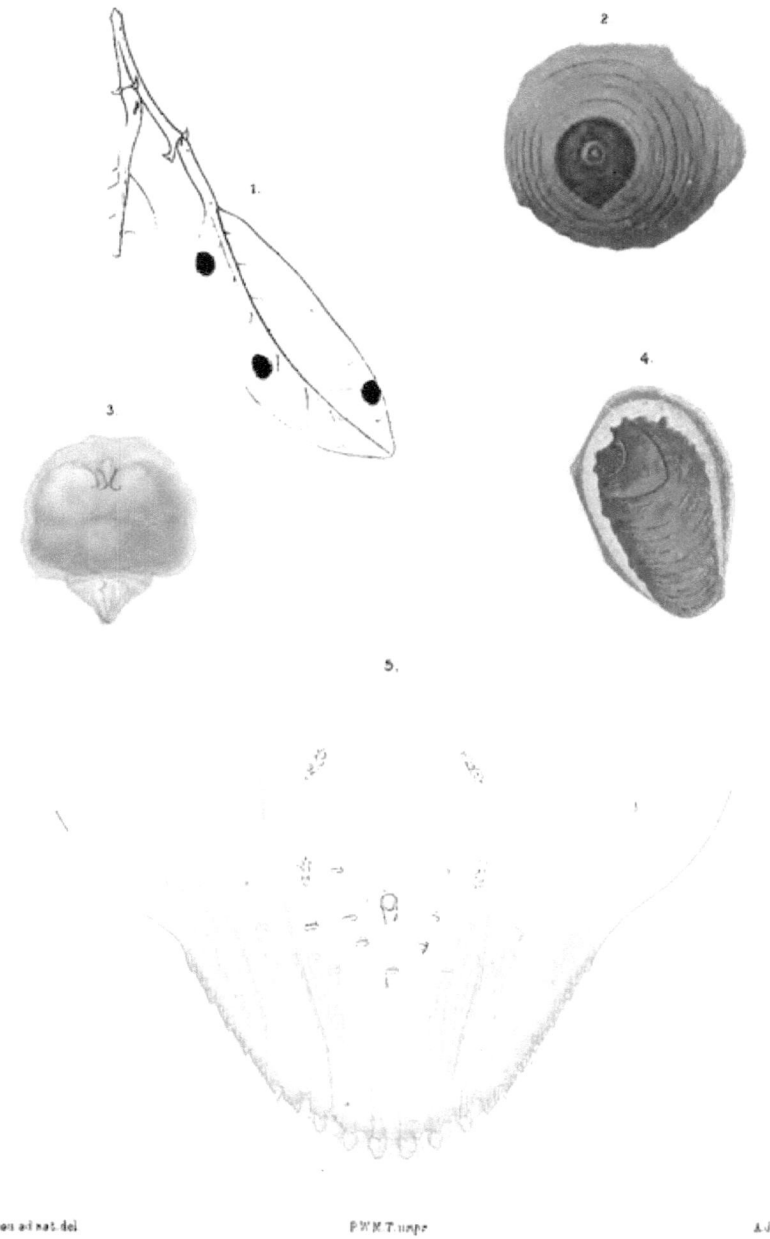

ASPIDIOTUS ROSSI.

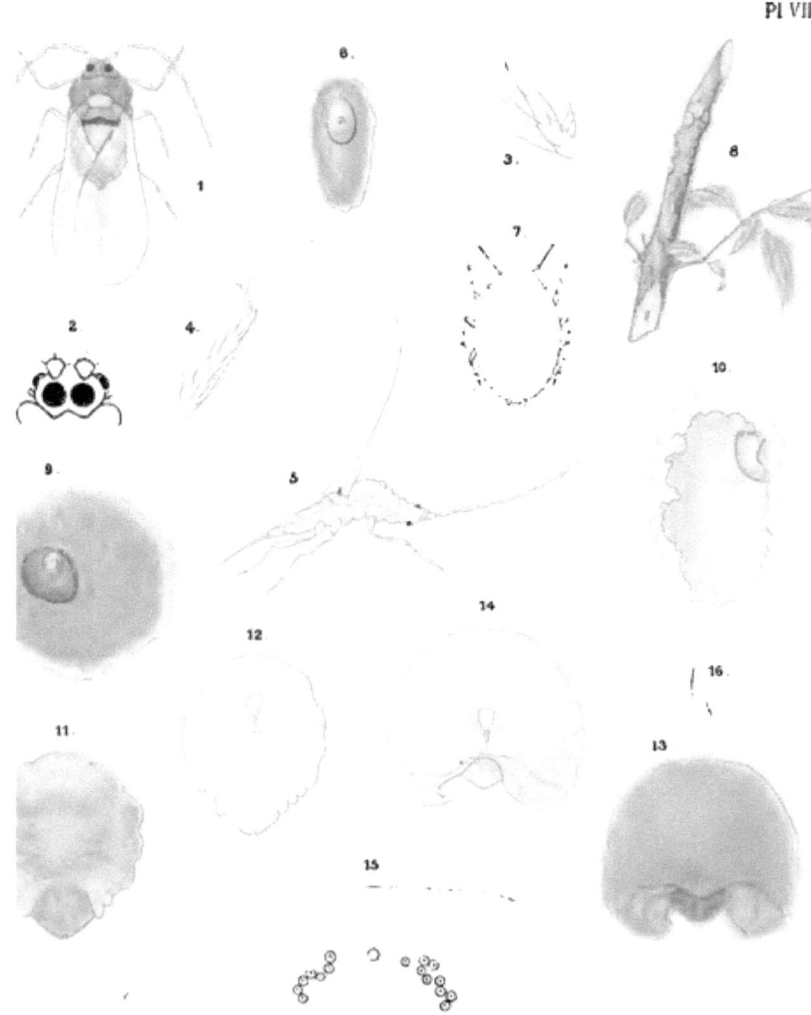

Pl VII

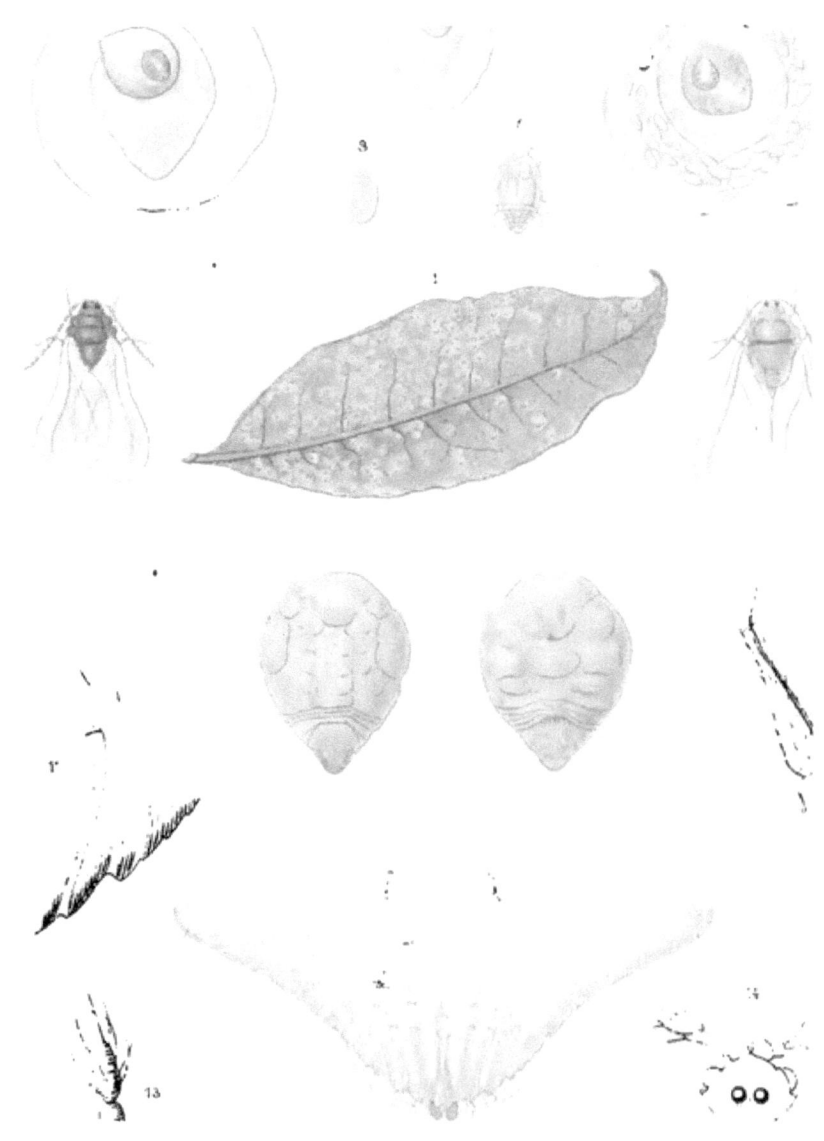

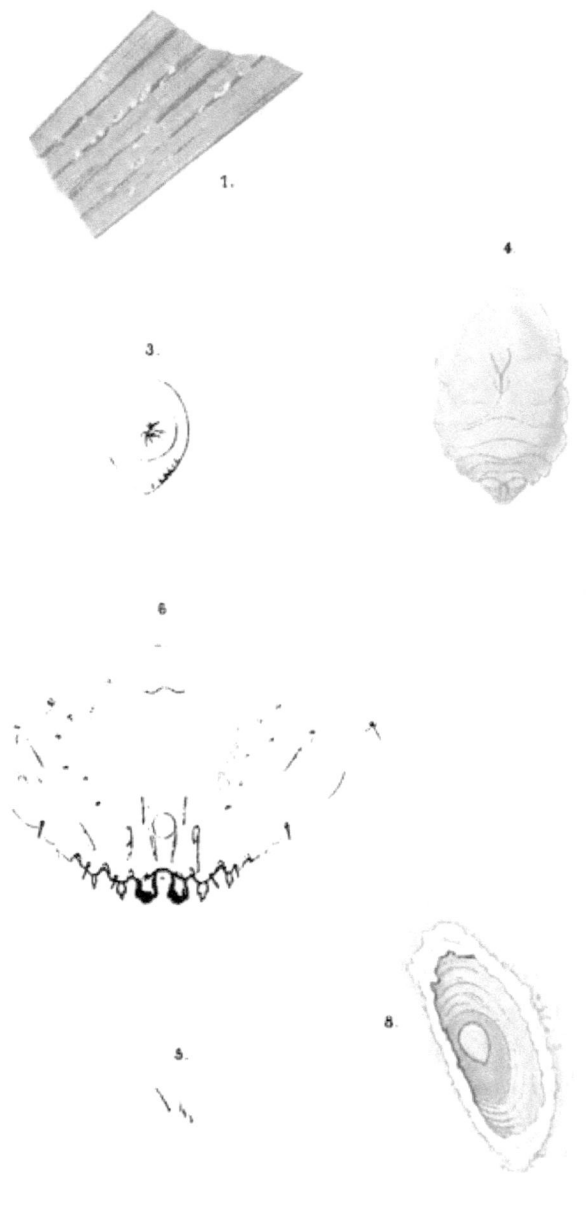

ASPIDIOTUS CYANOPHYLLI.

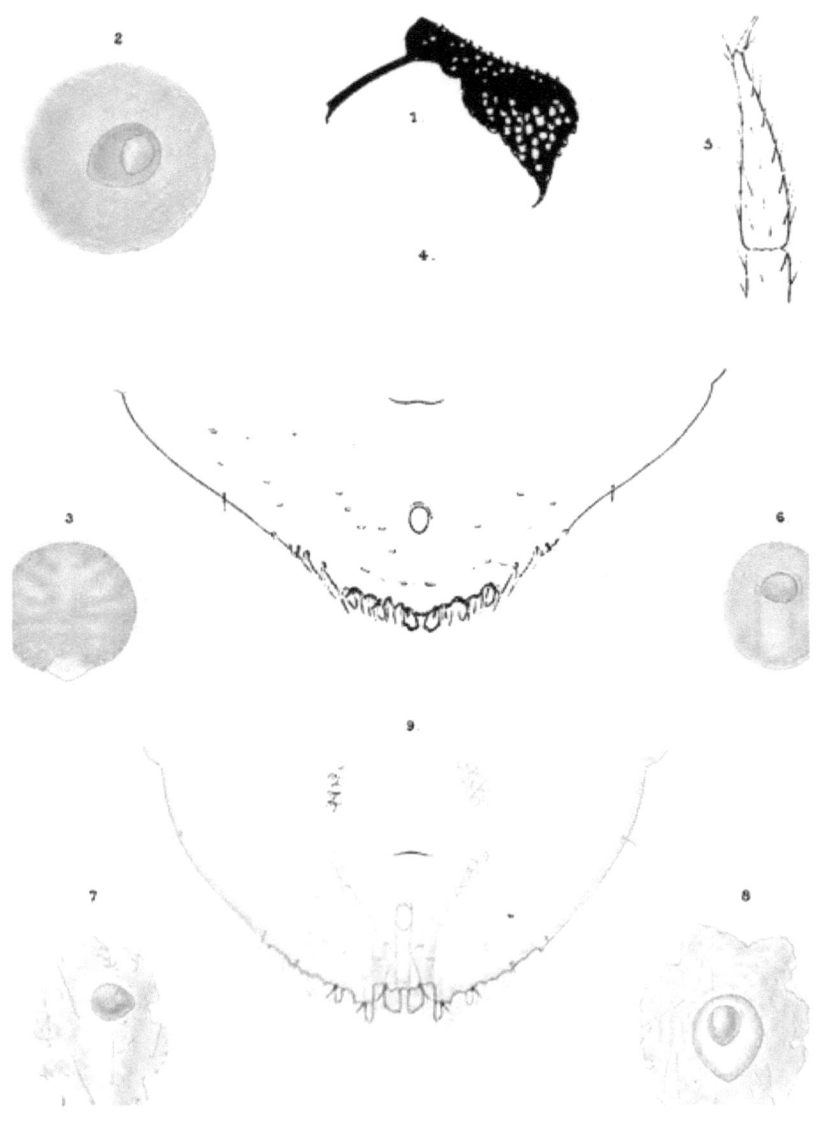

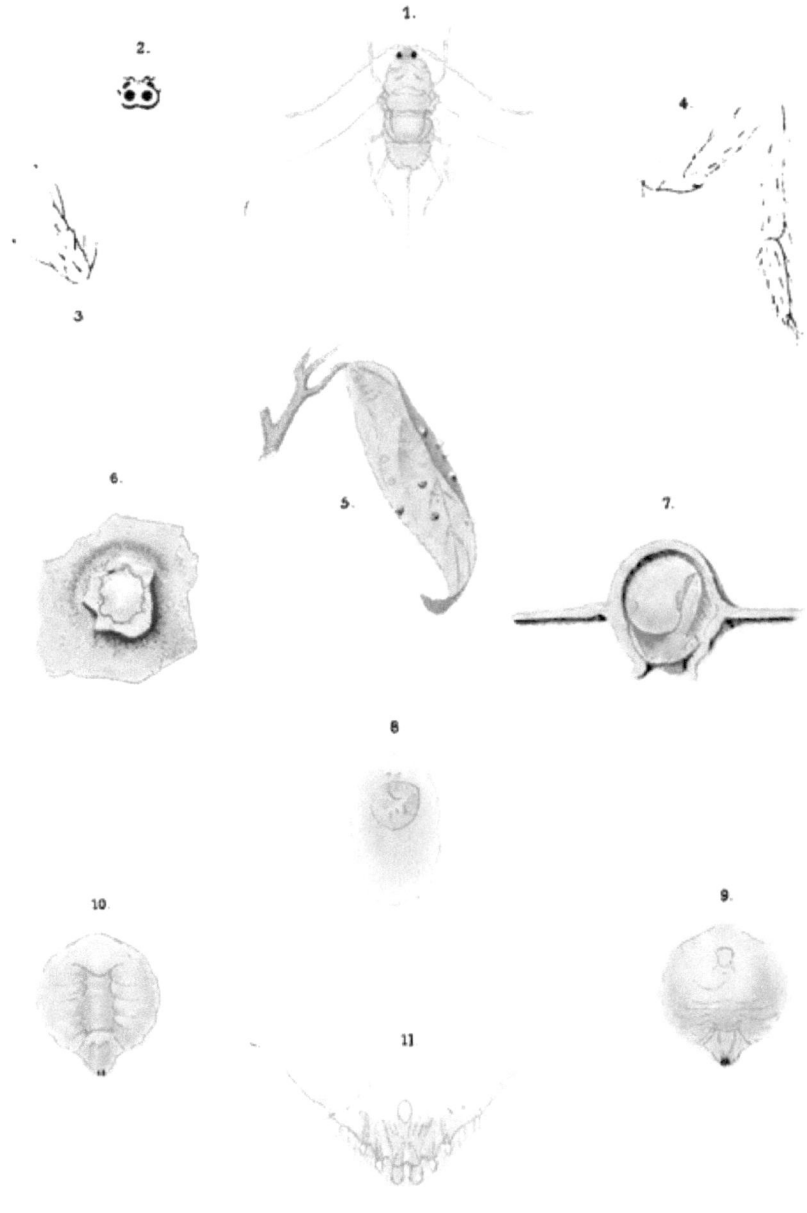

ASPIDIOTUS OCCULTUS.

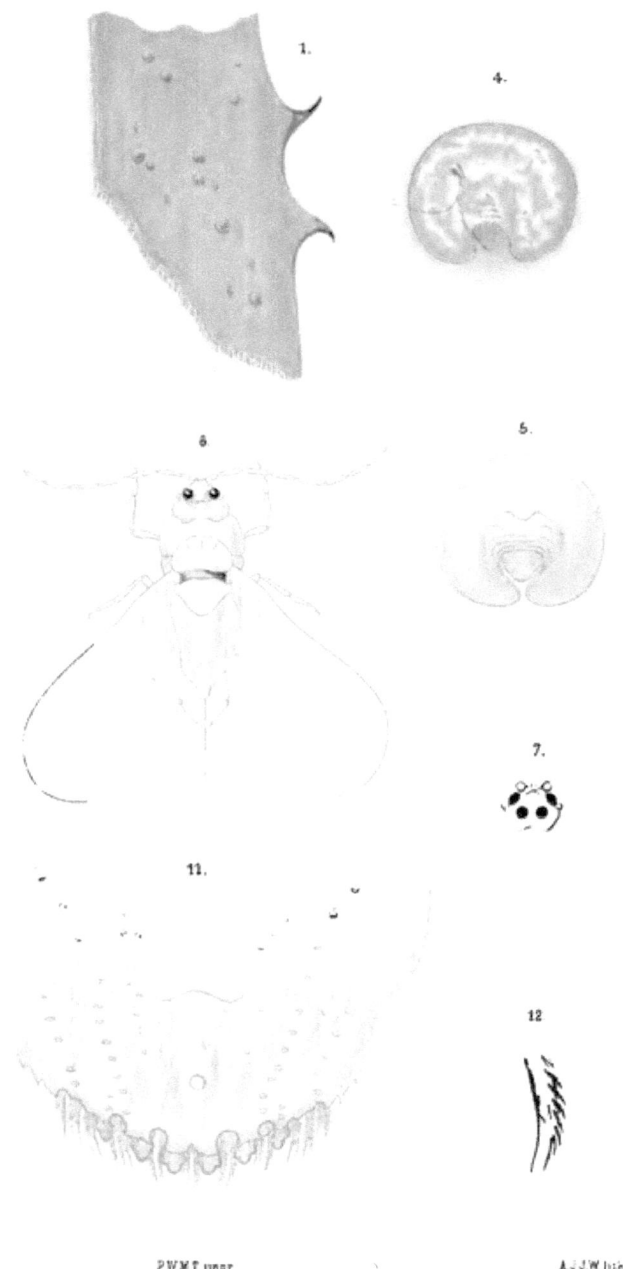

ASPIDIOTUS AURANTII.

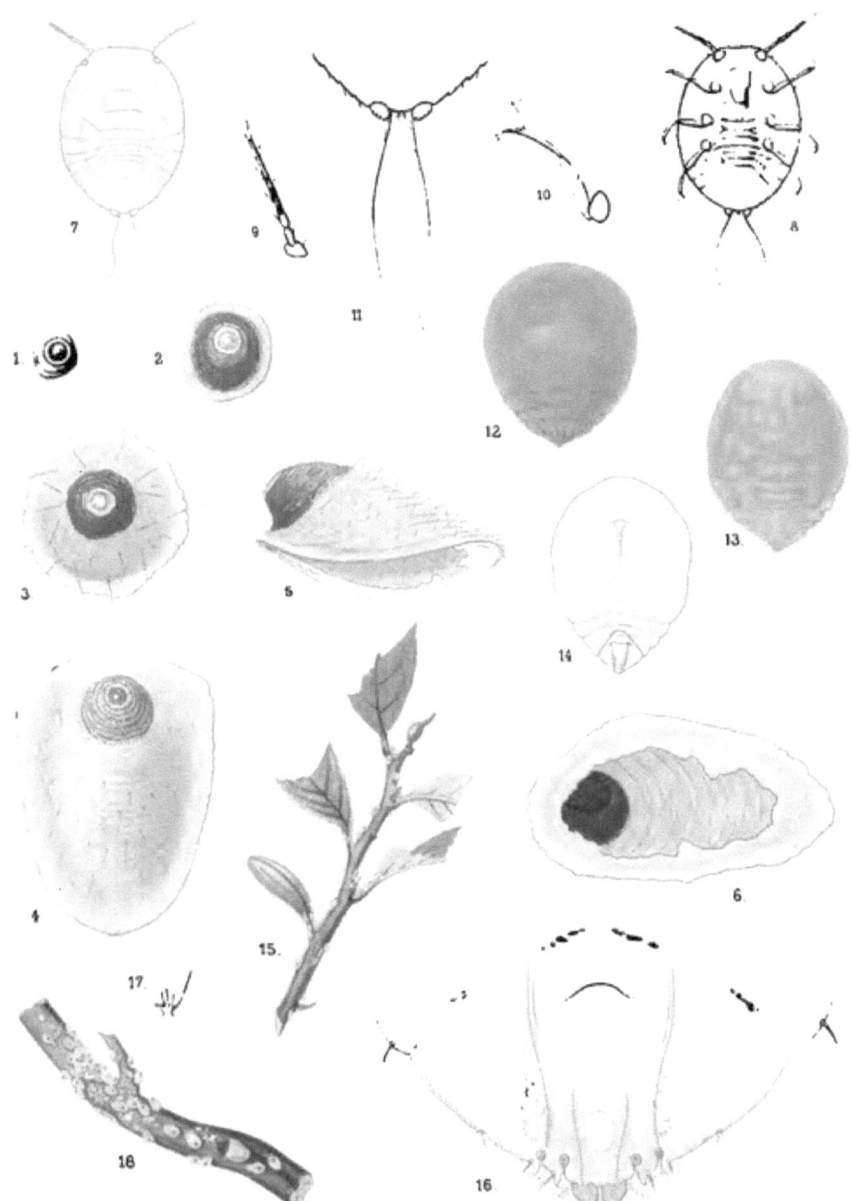

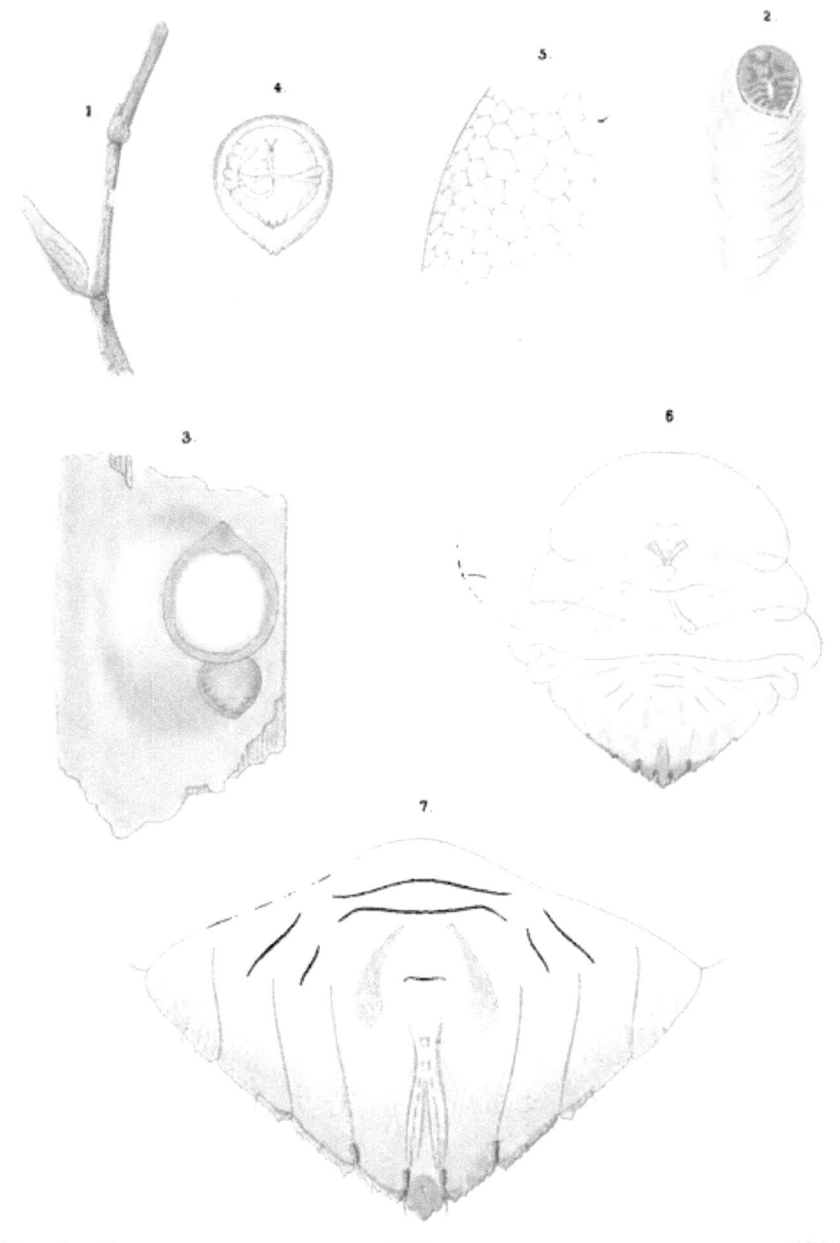

ASPIDIOTUS SECRETUS.

ASPIDIOTUS INUSITATUS.

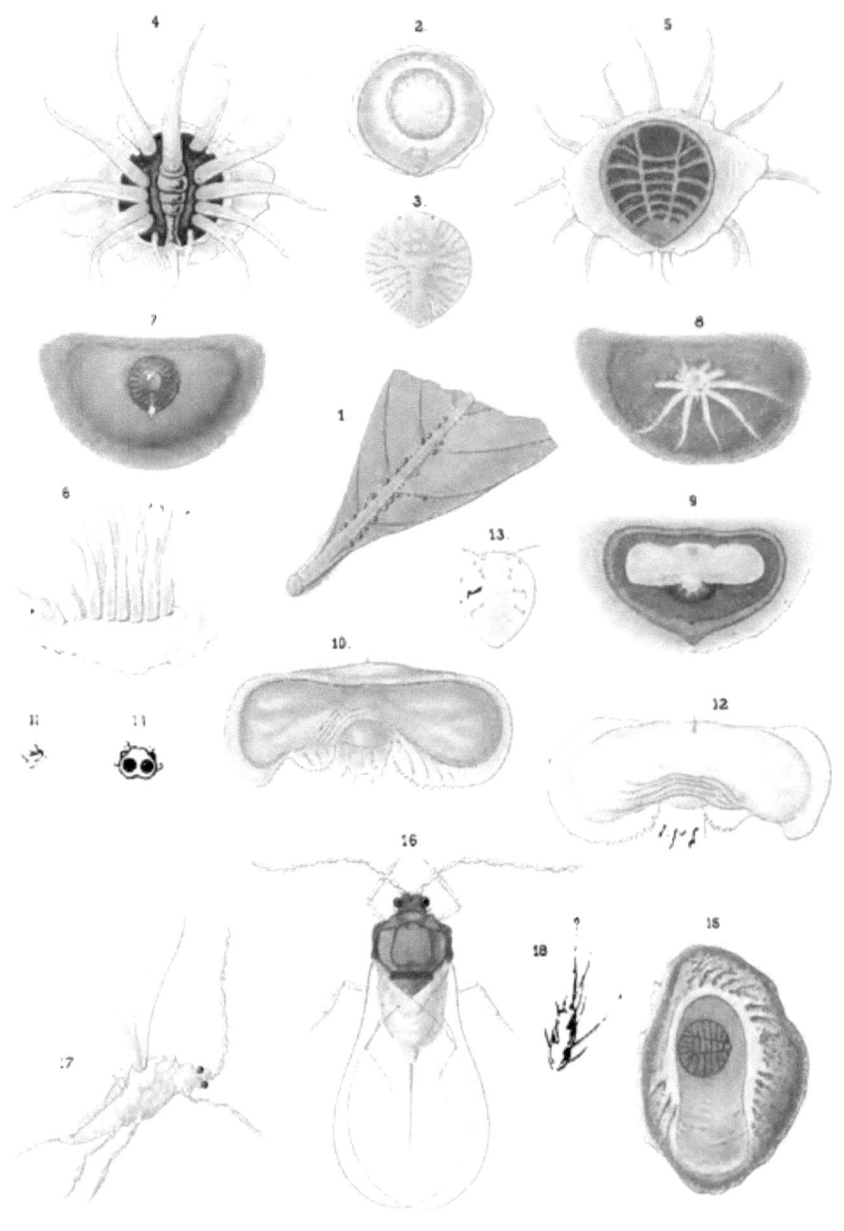

AONIDIA CORNIGER

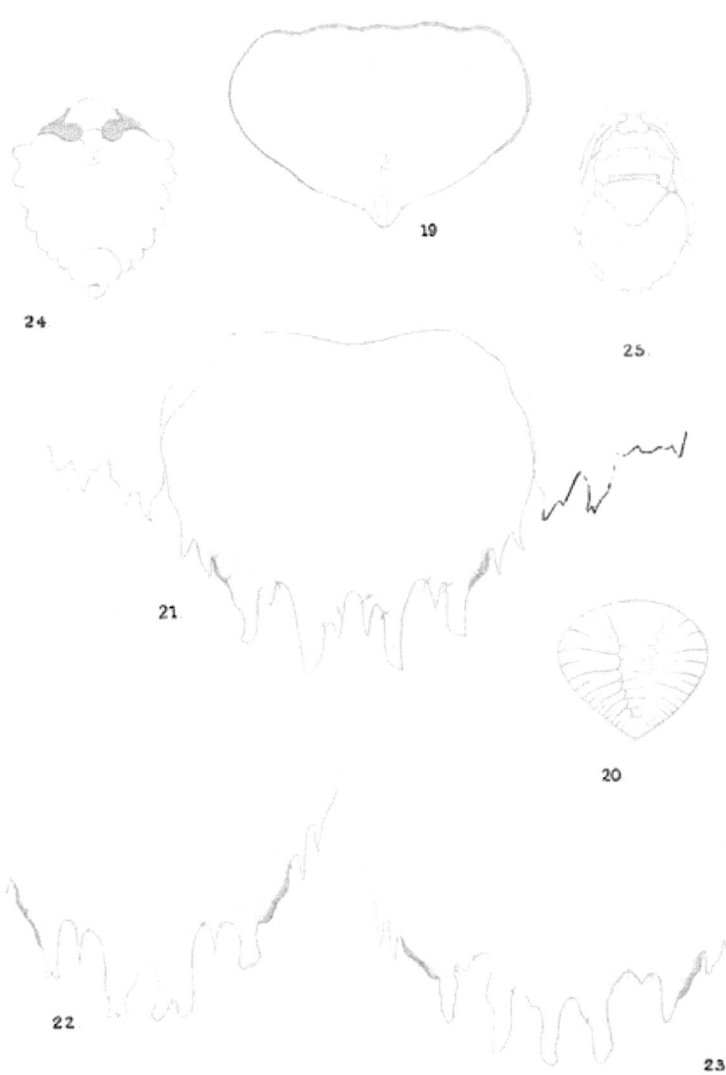

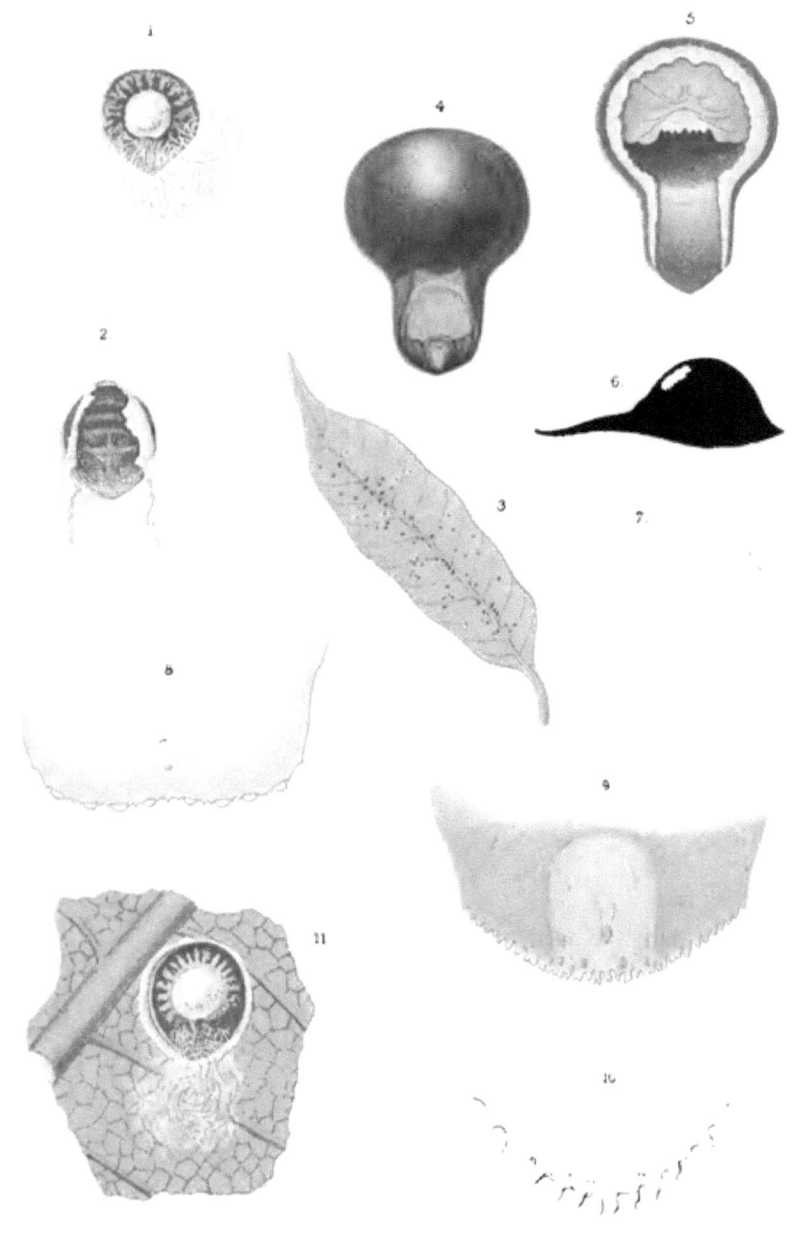

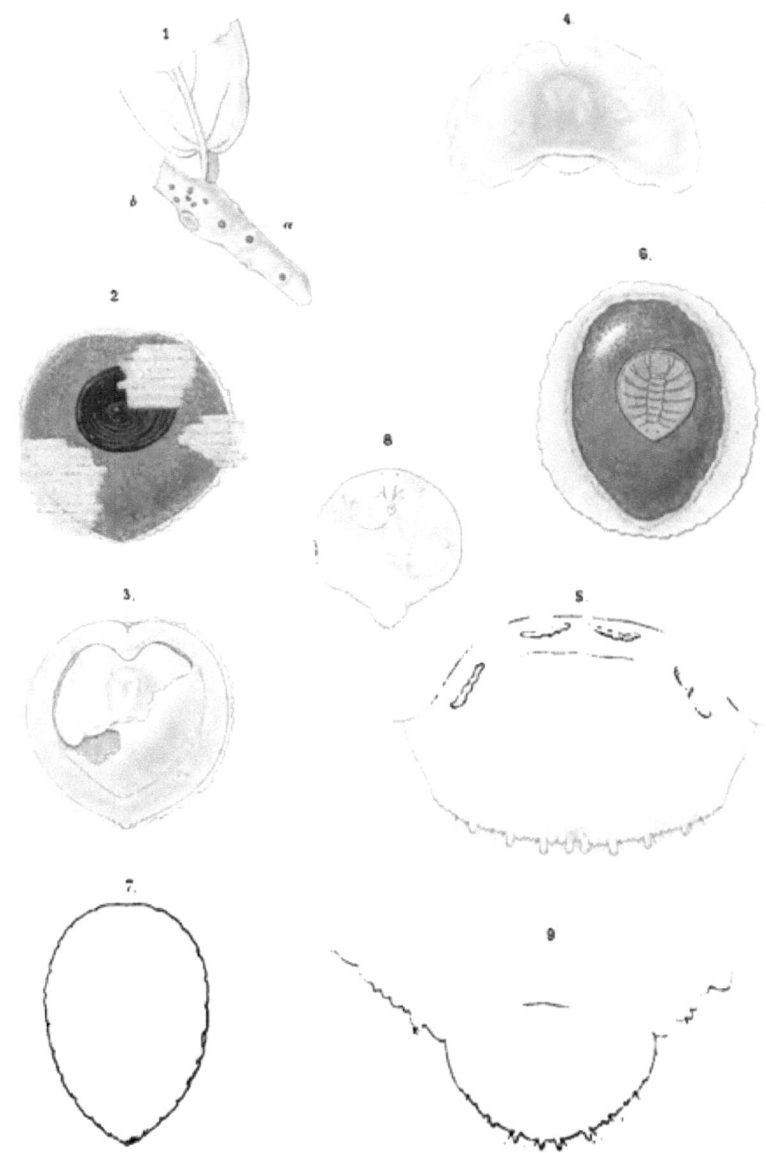

Pl. XIX

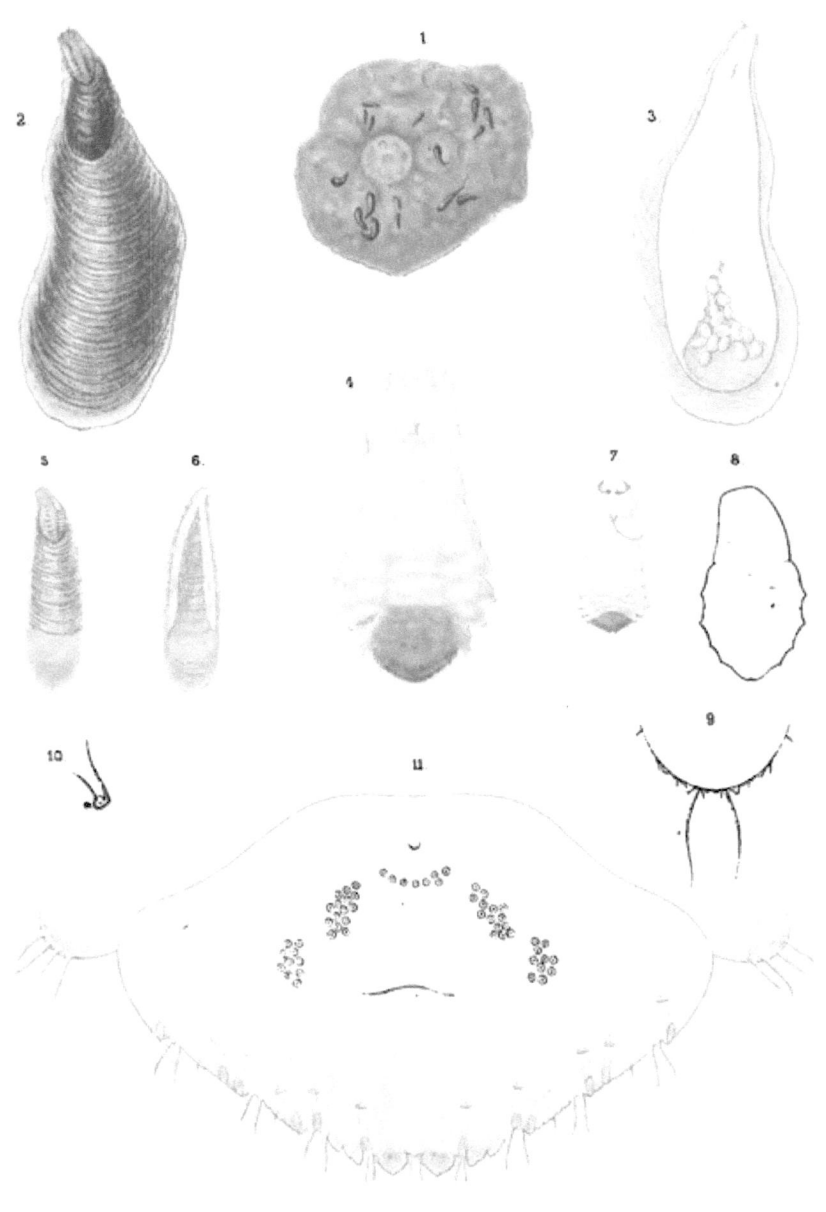

MYTILASPIS CITRICOLA.

Pl. XXI

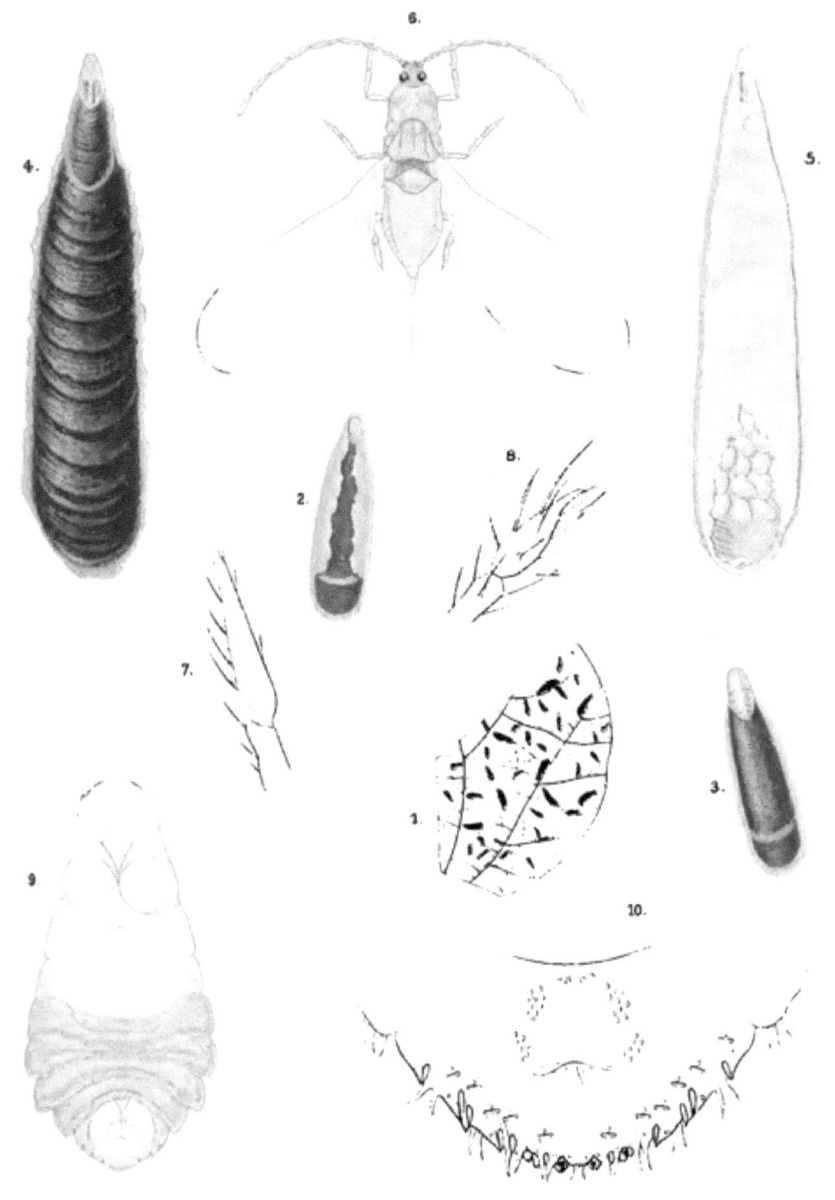

MYTILASPIS COCCULI

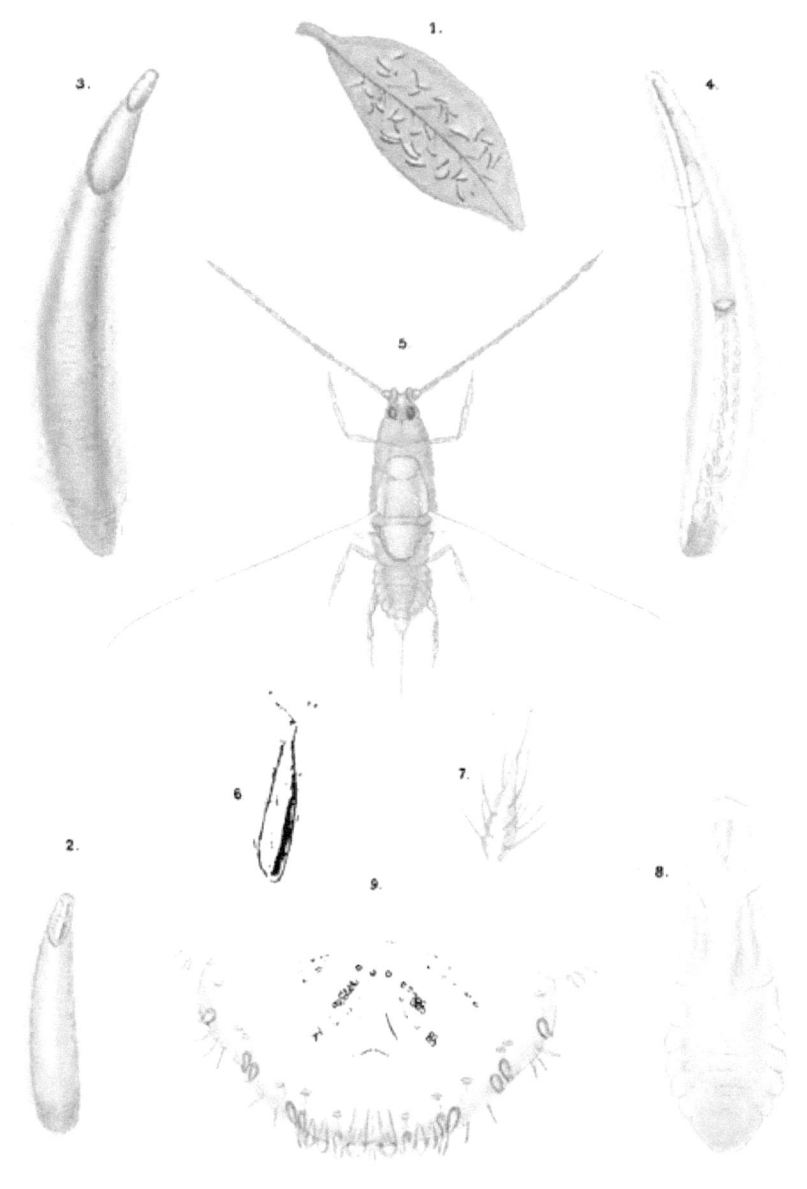

MYTILASPIS GLOVERII. VAR. PALLIDA.

Pl. XXIV.

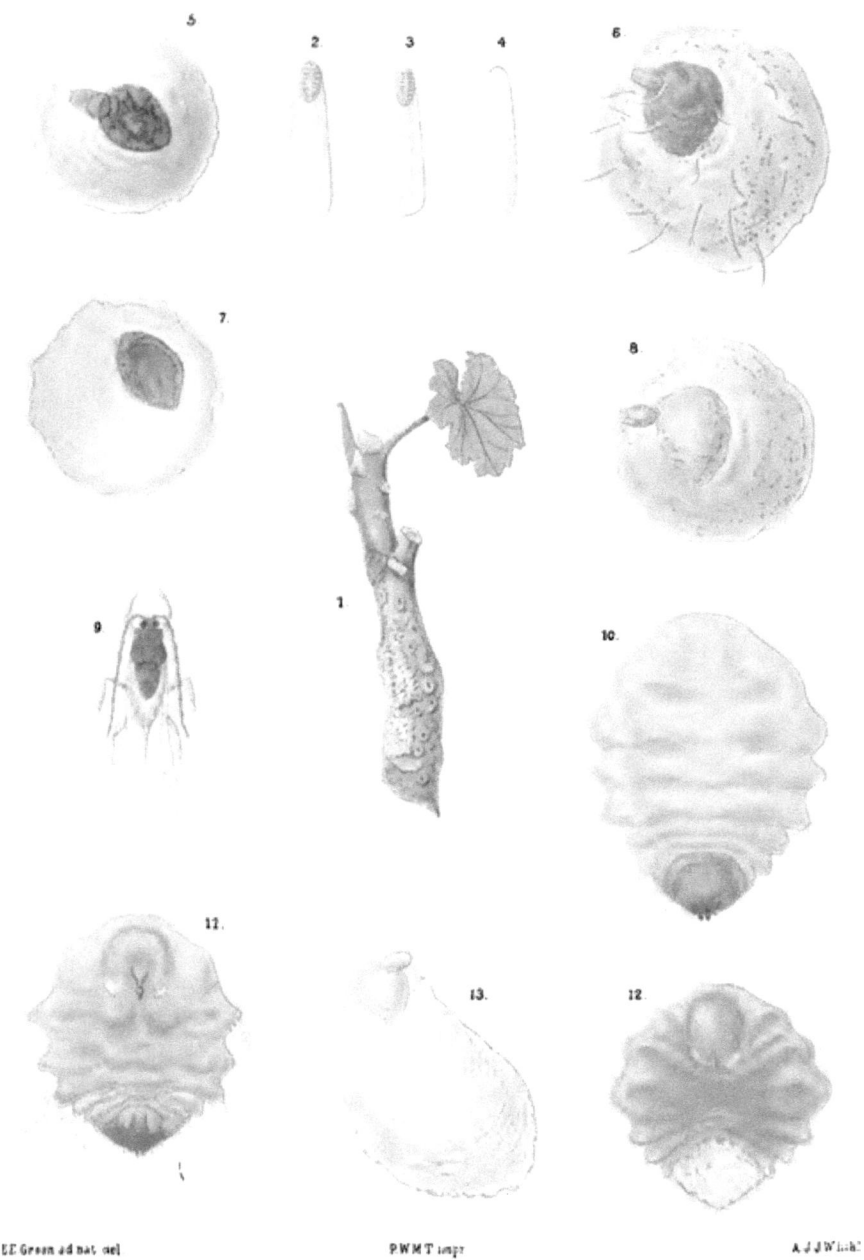

E. E. Green ad nat. del.   P.W.M.T. impr.   A.J.J.W. lith.

DIASPIS AMYGDALI.

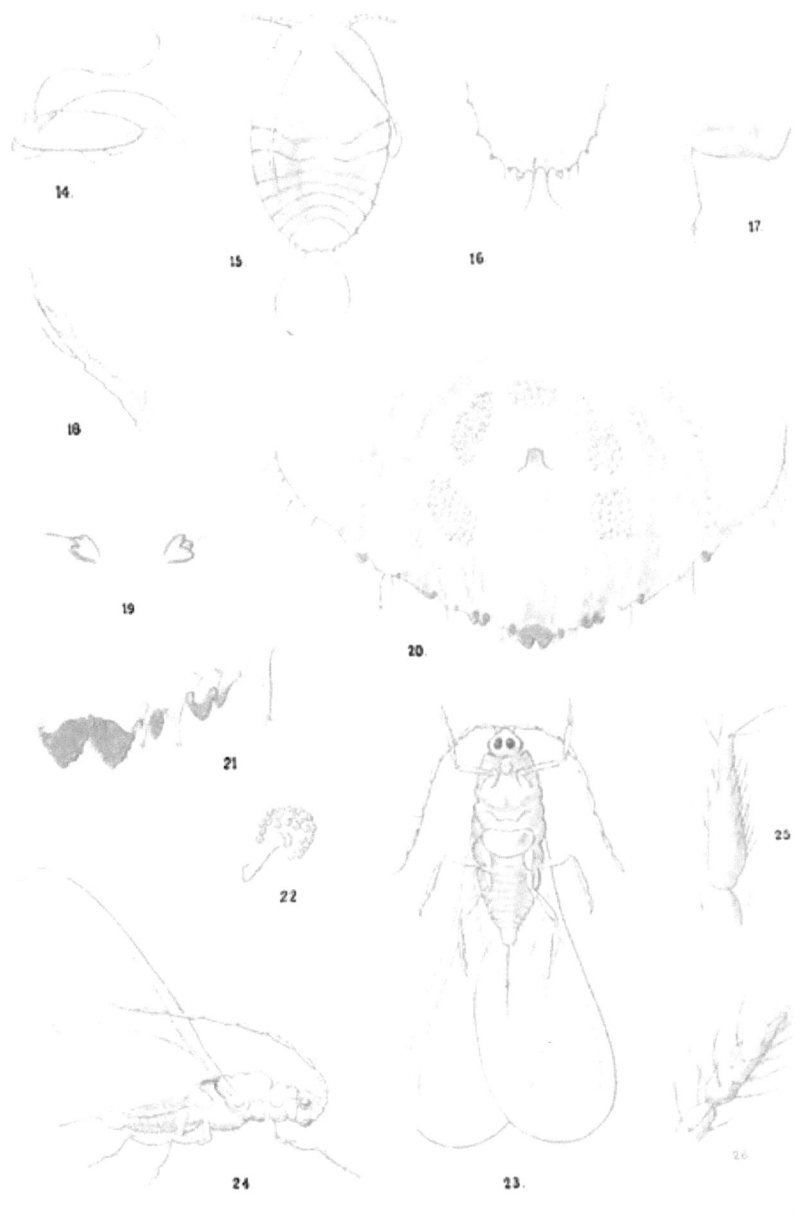

DIASPIS AMYGDALI.

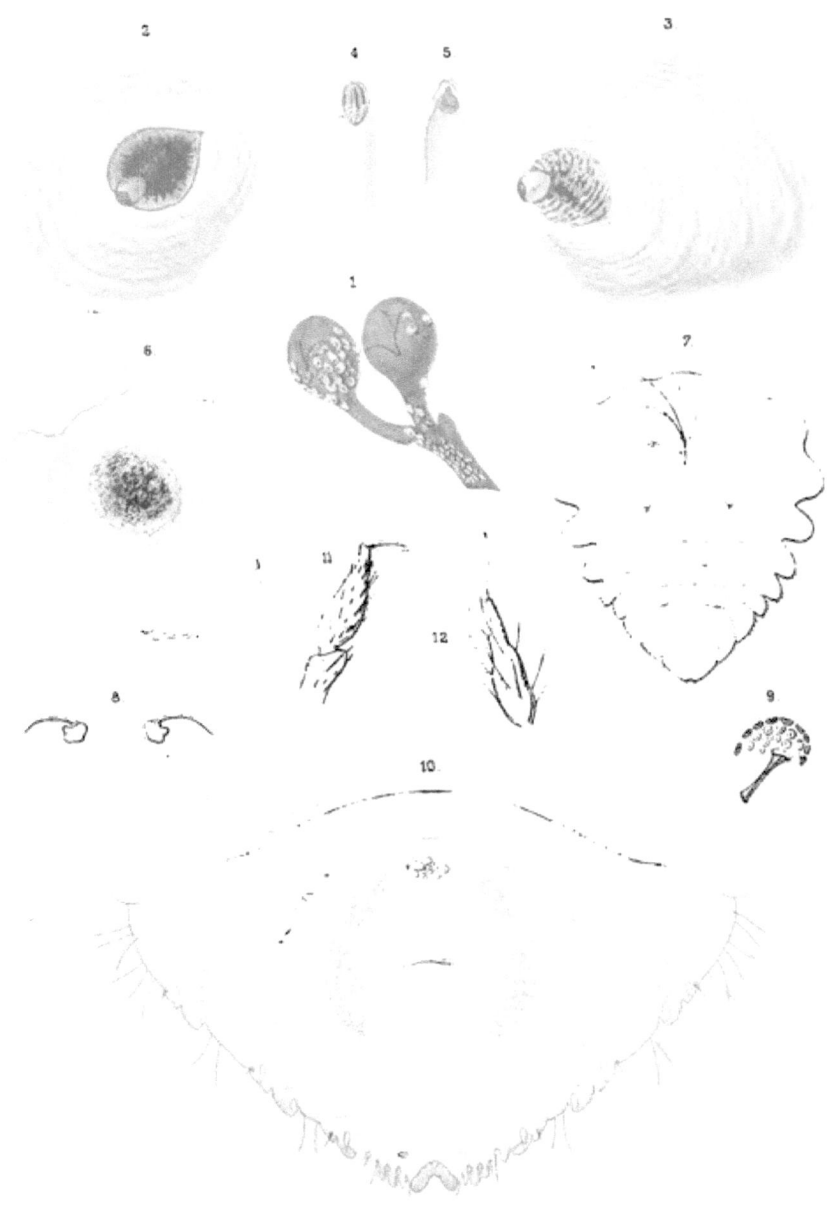

DIASPIS FAGRÆÆ.

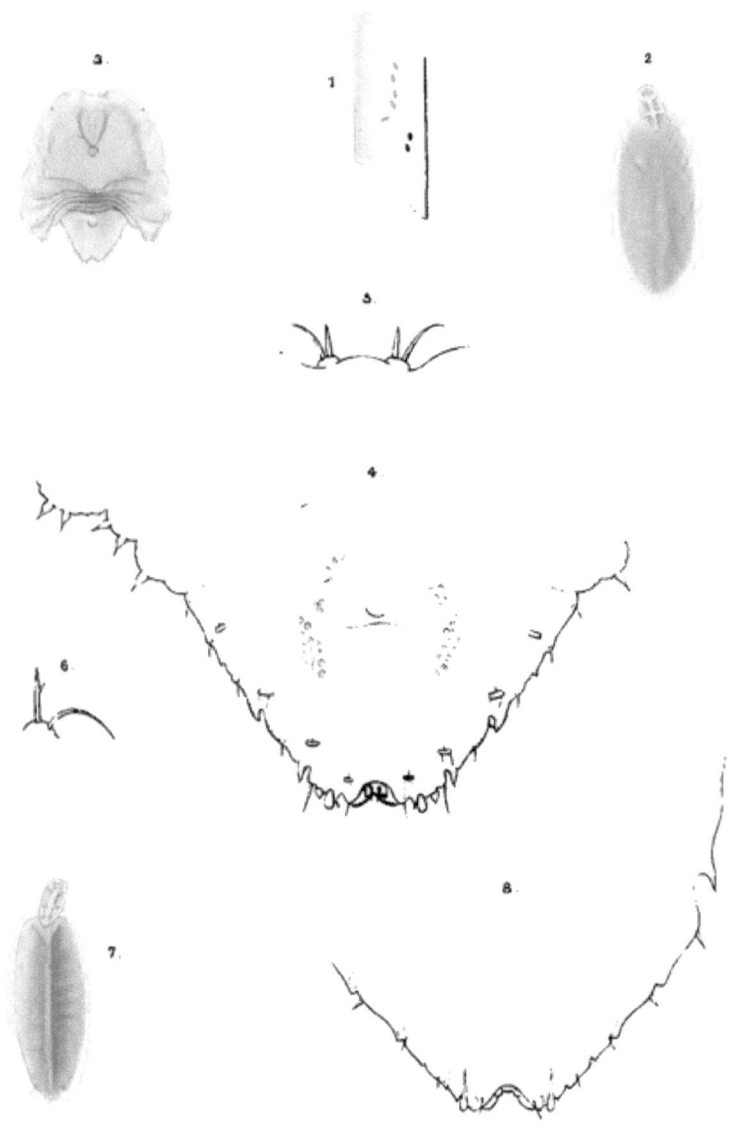

FIORINIA FIORINIÆ.

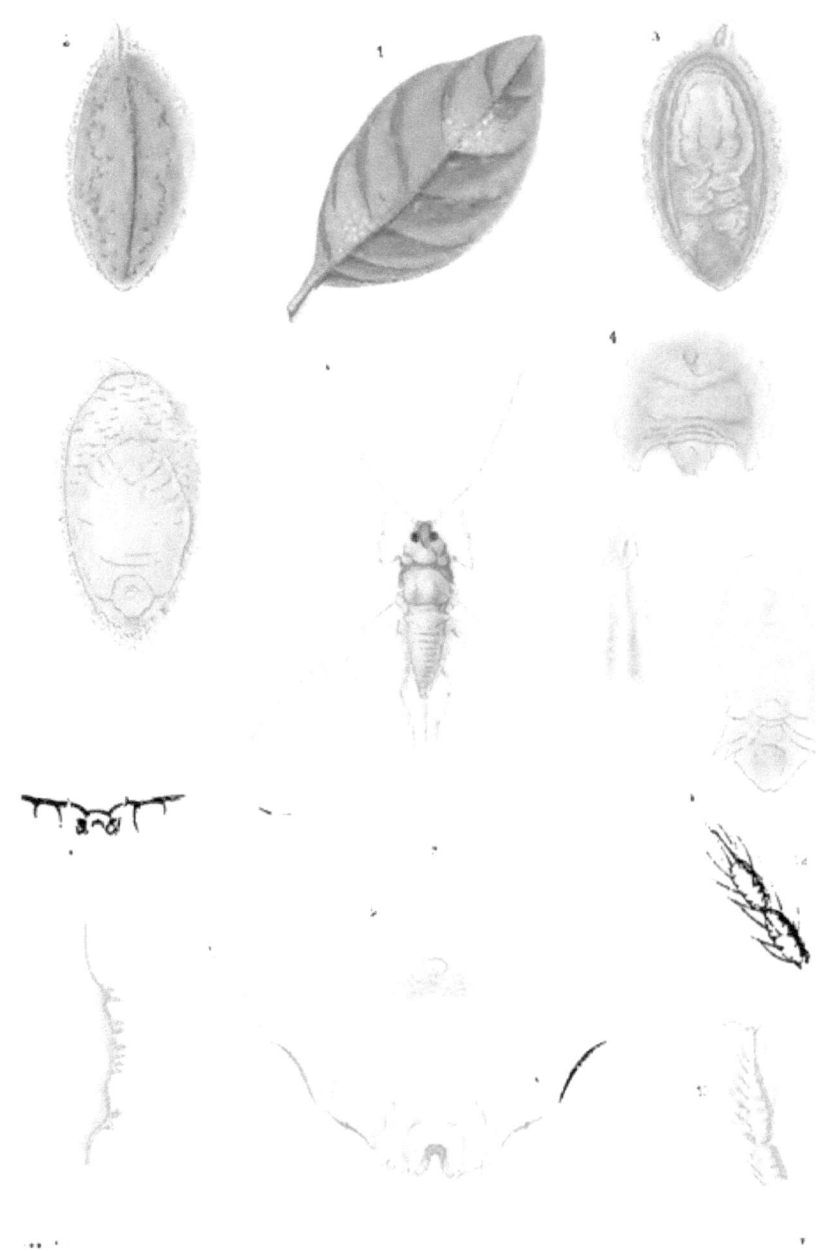

FIORINIA SAPROSMÆ.

Pl. XXVIII.

FIORINIA SIMILIS

www.ingramcontent.com/pod-product-compliance
Lightning Source LLC
Chambersburg PA
CBHW020249170426
43202CB00008B/284